The Open University

MU120
Open Mathematics

Resource Book C

Units 10–13

This publication forms part of an Open University course. Details of this and other Open University courses can be obtained from the Student Registration and Enquiry Service, The Open University, PO Box 197, Milton Keynes, MK7 6BJ, United Kingdom: tel. +44 (0)870 333 4340, e-mail general-enquiries@open.ac.uk

Alternatively, you may visit the Open University website at http://www.open.ac.uk where you can learn more about the wide range of courses and packs offered at all levels by The Open University.

To purchase a selection of Open University course materials, visit the webshop at www.ouw.co.uk, or contact Open University Worldwide, Michael Young Building, Walton Hall, Milton Keynes, MK7 6AA, United Kingdom, for a brochure: tel. +44 (0)1908 858785, fax +44 (0)1908 858787, e-mail ouwenq@open.ac.uk

The Open University, Walton Hall, Milton Keynes, MK7 6AA.

First published 1996. Reprinted 2005, 2006.

Edited, designed and typeset by The Open University, using the Open University TEX System.

Printed and bound in the United Kingdom by The Charlesworth Group, Wakefield.

ISBN 0 7492 5778 4

2.3

Contents

Unit 10 Prediction

Question 1

As an Open University mathematics lecturer, I have a pigeon-hole here at Walton Hall. The pigeon-hole is 10 cm high and I estimate that I get on average 2 cm of mail every week. When I am away on study leave, I ask my secretary to empty it, so that it does not overflow. Just to make sure, she likes to empty it when there is 1 cm of space left. She wants to know how often she will need to empty the pigeon-hole.

(a) Draw a linear graph to show how the height of the pile of mail, h cm, changes with time, t weeks.

(b) What is the question you want to answer from analysing this graphical linear model?

(c) What is the answer to your question and what is its interpretation in terms of the original problem?

(d) Find the gradient of your graph. What is its physical significance?

Question 2

In this question you are asked to model the same situation as in the previous question but this time with a different choice of variables. Instead of choosing the height of the pile of mail and time, in this question use the amount of space left in the pigeon-hole, s cm, and time, t weeks. Initially, at time $t = 0$ weeks, there is $s = 10$ cm of space left in the pigeon-hole.

(a) Draw a linear graph to show how the space left in the pigeon-hole, s cm, changes with time, t weeks.

(b) What is the question you want to answer from analysing this graphical linear model?

(c) What is the answer to your question and what is its interpretation in terms of the original problem?

(d) Find the gradient of your graph. What is its physical significance?

Question 3

Late spring frost is a threat to the blossom in Californian orange groves, so fires are sometimes lit between the trees when a frost is imminent in order to keep the temperature above freezing point. Rather than stay up all night, the workers watch how the temperature is falling after sunset and estimate how low it will go during the night. They then get up and light the fires shortly before their estimate of when the temperature will reach freezing point. From experience they know that the danger of frost occurs on cold clear nights and that the temperature usually drops steadily between sunset and sunrise. These workers are unlikely to write down an algebraic equation but their method of predicting when to light fires is equivalent to an algebraic linear model.

Look at the situation on one particular night. The temperature was 5 °C at sunset (6.30 pm), 4.5 °C an hour later, and 4 °C after a further hour.

You are asked to model the situation to help the workers to decide when to get up and light their fires on this night.

(a) Define the variables.

(b) Assume a constant rate of change of temperature. Find a numerical value for it.

(c) Find an initial value.

(d) Write down the appropriate linear equation.

(e) What are the limitations of the model?

(f) How could you use this model to find at what time the temperature will drop to freezing point (0 °C)?

(g) Find the solution and interpret it in terms of the workers' problem.

Question 4

Look at the following data, which were obtained by suspending various known masses on an elastic band and measuring the corresponding lengths of the elastic band.

Mass, x (g)	100	200	300	400	500	600
Length, y (mm)	228	236	256	278	285	301

(a) Plot the data pairs in the table as points on a graph. Draw 'by eye' the straight line which you think is the 'best fit' to the points you have plotted.

(b) Use your straight-line graph to estimate:

(i) the length of the elastic band before it was stretched—that is, the initial value;

(ii) the increase in the length of the elastic band caused by an additional mass of 1 g—that is, the gradient of the graph.

(c) By using your answers to part (b), write down the equation of your 'best fit' line.

(d) Now enter the data on your calculator and use it to find the equation of the regression line. Estimate the length of the band correct to the nearest mm when it has a mass suspended from it of:

(i) 350 g; (ii) 800 g; (iii) 1200 g.

Question 5

A slow goods train is scheduled to leave Edinburgh at 10.30 am. An additional holiday train is scheduled to leave half an hour later, and follows the same line to Newcastle (about 125 miles away). The holiday train travels faster than the goods train, and so the timetabling office want to known where to pull the goods train off the main line in order to let the holiday train go through. The goods train averages about 50 mph while the holiday train averages about 90 mph.

(a) On paper, sketch a graphical linear model of the motion of the goods train. On the same sketch construct a graphical linear model of the motion of the holiday train.

(b) By entering appropriate equations on your calculator, draw the graphs of the two models and so predict where the trains will meet if the goods train is not pulled off the main line.

(c) Interpret your solution in terms of the original problem.

Question 6

Suppose you were crossing the English Channel from Folkestone to Calais, a trip of 30 miles. Your ferry left at 20.50 and was scheduled to arrive at 22.40. From the timetable you noticed that another ferry was due to leave Calais at 20.15 and arrive at Folkestone at 22.05, and you were curious as to when you would pass this ferry. Sketch graphical linear models of the journeys of the two ferries. Enter appropriate equations in your calculator and use them to predict when you should look out for the other ferry.

Question 7

Solve the following pairs of simultaneous equations for the two unknowns x and y. For this question, use an algebraic approach rather than plotting the equations on your calculator.

(a) $y = 3 + x$
$y = 3x + 7$

(b) $y = 2 - 3x$
$3y = {}^{-}4 - 4x$

(c) $x = y - 1$
$3y = 2x - 2$

(d) $y = 3 - 2x$
$y = 2x - 1$

(e) $y = 10 + 5x$
$2y = 3x + 1$

(f) $2y + x = 7$
$3y + 2x = 4$

Question 8

Look back at Activity 27 of *Unit 10*, involving a graphical model of the supply and demand of soft fruit.

(a) From your graphs, find linear equations for the supply and demand in terms of the price.

(b) Use algebraic methods to solve these equations to find the equilibrium price, when the quantity of soft fruit demanded is exactly the same as the quantity supplied.

(c) How much fruit will be sold at this price according to your model?

(d) What assumptions were made in this model, and how do these affect the interpretation of the solution?

Question 9

For each of the following sets of conditions, draw a graph and mark on it the region where all of the conditions hold true.

(a) $x \geq 0, \quad y \geq 0, \quad 2x + 3y \leq 60.$

(b) $x + y \leq 2, \quad x + y \geq {}^{-}2, \quad x - y \geq -2, \quad x - y \leq 2.$

(c) $x \leq 50, \quad x/6 + y/8 \leq 10, \quad y + 4x \geq 80, \quad 2y + x \geq 60.$

Question 10

(a) Sketch the feasible region determined by the following set of inequalities and work out the coordinates of the vertices of the region.

$$x \geq 0, \quad y \geq 0, \quad 2x + y \leq 8, \quad x + 3y \leq 9.$$

(b) Find graphically the point of the feasible region in (a) which maximizes P in the cases:

(i) $P = x + y$;

(ii) $P = 3x + y$.

Question 11

(a) Sketch the feasible region determined by the following set of inequalities and work out the coordinates of the vertices of the region.

$$x \geq 0, \quad y \geq 0, \quad x + 2y \leq 12, \quad 5x + y \leq 15, \quad 4x + 3y \leq 30.$$

(b) Find graphically the point of the feasible region in (a) which maximizes P in the cases:

(i) $P = x + y$;

(ii) $P = 7x + 2y$.

Question 12

Suppose that you have agreed to bake some cakes for a party at a club you belong to. The club will provide all the necessary ingredients and a fully equipped kitchen, and will also pay you 80p each for fruit cakes and £1 each for iced cherry cakes. Suppose you estimate that fruit cakes take one hour of your time to prepare, and $1\frac{1}{2}$ hours in the oven; and that iced cakes take two hours of your time but only use one hour of oven time. You have 18 hours of time available for preparation, and 14 hours of time to use the oven (the oven is only big enough for one cake at a time).

(a) Model this problem as a linear programming problem. Sketch the feasible region, and hence determine how many cakes of each type you should make in order to maximize the income you will receive.

(b) Suppose the price you receive for iced cakes increases to £1.50 each. Does this change your answer to (a)?

(c) What would happen if the price for iced cakes rose to £1.70 each?

Question 13

Suppose that you have been reading about a new wonder diet in which you eat only bread and low-fat cheese. The nutritional details for a slimmers' bread and a low-fat cheese are given in the following table.

	Bread	Cheese	Required daily intake (g)
Protein (%)	8	16	75
Carbohydrates (%)	46	6	250
Fat (%)	2	8	70
Calories per 100 g	235	166	

The table also shows the required minimum daily intake of protein, carbohydrates and fat for a healthy diet. (Of course, a healthy diet also requires vitamins, minerals, and so on—but these are ignored here or you would not be able to solve the problem without using a computer.) How much bread and cheese should you eat each day so that your daily diet is healthy as far as protein, carbohydrates and fat are concerned, while at the same time minimizing your daily calorie intake?

(a) Model this problem as a linear programming problem, using as variables your daily intake in 100 g units of bread (x) and cheese (y), and draw the feasible region.

(b) What can you say about the protein requirement?

(c) Use your diagram or calculator to find the solution to your problem graphically.

(d) Find the optimal solution using algebra to solve the appropriate simultaneous equations.

(e) How many calories per day will you be getting?

Unit 11 Movement

Question 1

(a) Write down the general equation of a parabola with its axis parallel to the y-axis and with its vertex at:

(i) $(0, 4)$; (ii) $(4, 0)$; (iii) $(1, {}^{-}2)$.

(b) Write down the equations of two parabolas with axes parallel to the y-axis and with vertex at $(6, {}^{-}4)$.

Question 2

Write down the general equation of the parabola with axis parallel to the y-axis, vertex at $({}^{-}1, 2)$, and which goes through the point $(1, 14)$.

Question 3

Write down the coordinates of the vertex of each of the following parabolas.

(a) $y = (x + 1)^2 + 1$

(b) $y = x^2 + 1$

(c) $y = (x - 2)^2 + 1$

(d) $y = 2(x - 3)^2 + 2$

Question 4

Suppose that a marine biologist is researching on certain type of (very fast growing) coral. An experiment is performed on a piece of coral by weighing it every month. The data are collected in the table below.

Time (months)	1	2	3	4	5	6	7	8	9	10	11
Weight (g)	1580	2040	2880	3330	4590	6210	6430	8400	9930	9960	12380

(a) Enter the data into your calculator and draw a scatterplot of the data. Do you think that it is reasonable to fit a straight line to the data?

(b) Use the linear regression features of your calculator to fit a straight line to the data. Use the regression line to predict what the coral will weigh in month 15.

(c) The marine biologist has a theory that the growth of the coral is purely at the surface and so is proportional to the surface area of the coral. This suggests that a quadratic model is more appropriate for the data. Use your calculator to fit a quadratic model to the data, and use it to predict the weight of the coral in month 15.

(d) Comment on any differences between your predictions for the weight of the coral at 15 months.

Question 5

Suppose that an experimenter wants to analyse the motion of a stone as it is thrown from a cliff. The experimenter starts a watch and then throws the stone from the cliff. The experiment is filmed and the data from the experiment are presented as the graph in Figure 1.

Note that the height of the stone has been plotted against *time*, not against horizontal distance. This graph does not represent a picture of what you would see if you were viewing the motion from the side.

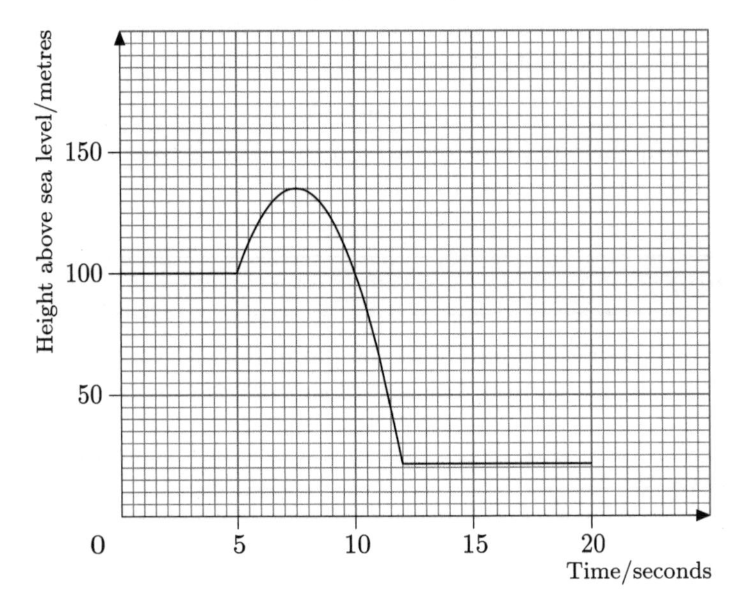

Figure 1

(a) What is the height above sea level of the top of the cliff?

(b) How long after starting his watch did the experimenter throw the stone?

(c) When did the stone return to a height level with the cliff edge?

(d) At what time did the stone reach its maximum height? How high did the stone go?

(e) Using the answers to the previous three parts of the question to define three points on the parabola, use your calculator to find the equation of the best fit parabola which goes through these three points.

(f) Given that the bottom of the cliff is 21 m above sea level, use your calculator to find the time at which the stone hit the ground.

(g) Use your calculator to calculate how fast the stone was travelling when it hit the ground (that is, find the gradient of the parabola when the stone hit the ground).

Question 6

A good model of certain towns and villages is a roughly circular shape. Sometimes the population density along a diameter can be modelled as a parabola. The population density along the diameter of a town modelled in this way is shown in Figure 2.

(a) Calculate the equation of the parabola that the graph represents (either algebraically or by fitting a parabola using your calculator).

(b) Use your equation to predict the population density 200 m from the centre of the town.

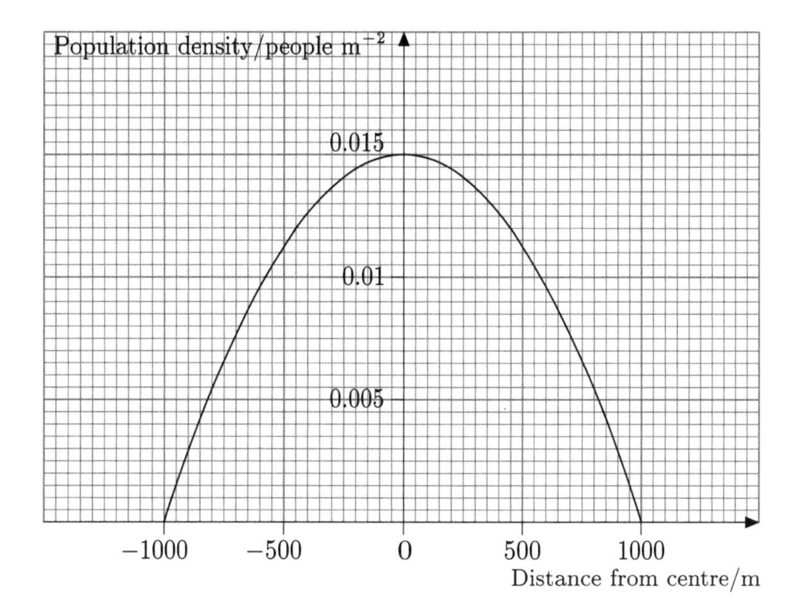

Figure 2

Question 7

Although many towns are roughly circular in shape, some villages have evolved along a main street in such a way that their population density has its peak value alongside the road and falls away each side of the road, as illustrated by Figure 3.

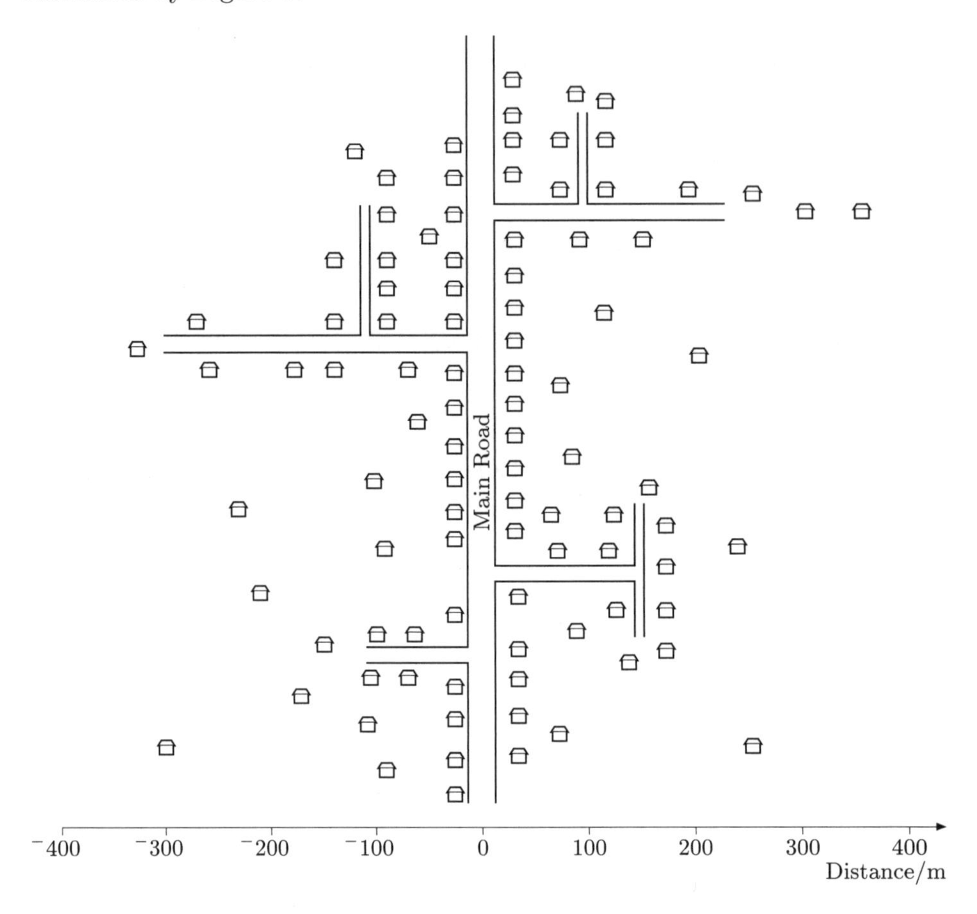

Figure 3

(a) Sketch how you think the average population density is likely to vary in relation to distance from the main road in Figure 3. What curve could model this variation? What would you have to assume about the population density outside the village boundary (400 m from the main road)?

(b) The population density is roughly 0.01 people m^{-2} alongside the main road. Make some appropriate modelling assumptions and use them and the given data to model the variation of population density with distance from the main road. What population density would you predict 200 m from the main road?

Question 8

(a) Solve the following quadratic equations by finding the square root. Give your answers correct to 2 decimal places.

(i) $x^2 = 36$

(ii) $x^2 = 30$

(iii) $5x^2 = 24$

(iv) $(x + 3)^2 = 18$

(b) Solve the following quadratic equations by factorization.

(i) $4x^2 - 6x = 0$

(ii) $x^2 + 3x - 18 = 0$

(iii) $2a^2 - 7a + 6 = 0$

(c) Where possible, solve the following quadratic equations by using the quadratic formula.

(i) $x^2 - 4x - 3 = 0$

(ii) $2x^2 - 7x + 4 = 0$

(iii) $3x^2 - 4x + 5 = 0$

Unit 12 Growth and decay

Question 1

In each of the following examples you are given the initial population and the growth factor. Calculate the population for the subsequent four stages or generations by direct calculation. Then devise a formula for the nth subsequent generation and use the formula to calculate the population in the 20th subsequent generation.

(a) A population of insects that starts with 5 members and triples every month.

(b) A fungus which adds 1% to its mass every day and starts off weighing 23.1 grams.

(c) A person who has just inherited £1000 and spends 10% of the remaining money each month.

Question 2

A modern day (and much less generous!) Queen Calcula offers to pay you 1 euro on the first day of the month, 1.01 euros on the second, 1.01^2 on the third and so on. Write down a general formula for the total paid over n days and use it to calculate the total amount paid during a 31-day month.

Question 3

(a) Using the method developed in Section 8.5 of the *Calculator Book*, use your calculator to solve graphically the equation $10^x = 20$, giving your answer correct to 3 decimal places.

(b) Use your calculator's log key to evaluate $\log_{10}(2)$, $\log_{10}(20)$ and $\log_{10}(200)$.

(c) Without using your calculator write down the value of $10^{3.30102996}$.

Question 4

For the following functions, calculate the population doubling times

(a) $y = 209 \times 3^x$

(b) $y = 0.43 \times (1.2)^x$

(c) $y = 65 \times (0.9)^x$

Question 5

Write down the equation of the exponential model in the form $y = a \times b^x$ for each of the following cases (round coefficients to four significant figures).

(a) Initial population 76, population doubling time 2 minutes.

(b) Initial population 10, population doubling time 6 minutes.

(c) Initial population 1000, half-life 23 minutes.

Question 6

The radioactive isotope uranium 239 has a half-life of 23.5 minutes.

(a) What is the 'quarter-life' of the isotope—that is, the time until the number of atoms of the isotope has decreased by a factor of four.

(b) Calculate the 'third-life' of the isotope—that is, the time until the number of atoms of the isotope has decreased by a factor of three.

Question 7

An environmental pressure group claims that the number of individuals in a certain species of bird is declining exponentially. The data that the group gives to support its claim is from an annual census project conducted over ten years. The data are collected together in the following table (the years are numbered from the start of the project; 1990 is year 0).

Year number	0	1	2	3	4	5	6	7	8	9
Population	515	481	441	390	339	307	275	227	195	164

(a) Fit an exponential model to the data as the pressure group suggests. Use your model to predict the population of birds in the year 2005.

(b) Now assume that the decline is linear and fit a linear model to the data. Use your linear model to make another preodiction of the population in the year 2005.

(c) Comment on your predictions for the population in the year 2005.

Question 8

(a) Two banks each offer a loan at the following rates:

Bank A 0.9% per month, compounded monthly.
Bank B 0.028% per day, compounded daily (assume 1 year = 365 days).

Convert both rates to APR and say which bank is offering the better deal.

(b) For an APR of 18%, calculate the corresponding rate of interest if it were calculated over the following periods:

(i) 1 month, compounded monthly;

(ii) 1 week, compounded weekly;

(iii) 1 day, compounded daily.

Give your answers correct to 3 decimal places.

Unit 13 Baker's dozen

Question 1

Suppose that it has taken you one hour to read the first twenty pages of a book and assume that you will read the rest at the same speed.

(a) Set up a mathematical model to help you predict how long it will take you to read any given number of pages of the book, by:

(i) choosing two appropriate variables;

(ii) making an assumption which leads to a proportionality relationship;

(iii) writing down this relationship, using the relevant constant of proportionality.

(b) Use the model to estimate how long it will take you to read:

(i) the first part of 115 pages;

(ii) the third part of 46 pages;

(iii) the whole book of 670 pages.

Question 2

Suppose that you were preparing a meal for some friends and were making a main course from a recipe which required a 15 cm diameter serving dish. At the last minute, when you had put all the ingredients together according to the recipe, you realized that your round serving dish had a diameter of 20 cm. How would the depth of the food in the serving dish be affected by using your dish instead of that given in the recipe (without increasing the ingredients)?

Question 3

Your recipe book lists the sizes for round cake tins but you have only got square cake tins. How do you convert from round to square (assuming you want the cakes to be the same height)? A particular recipe asks for an 8-inch diameter round cake tin. What is the corresponding length of side for a square tin of the same height?

Hint: the formula for the area of a square with side x inches is x^2 square inches; the formula for the area of a circle of radius r inches is πr^2 square inches.

Question 4

(a) The volume, V of a cylinder of radius r and height h is proportional to the square of the radius. The formula is as follows:

$$V = \pi r^2 h.$$

Rearrange the formula to find r in terms of V.

(b) The volume, V of a sphere of radius r is proportional to the cube of the radius. The formula is as follows:

$$V = \tfrac{4}{3}\pi r^3.$$

Express this relationship the other way round to find r in terms of V.

Question 5

You are swapping recipes with a friend and find that you can never make your friend's recipes work. Eventually you trace the problem to the different power ratings of your microwave ovens. You have a 650-watt microwave while your friend has a 500-watt microwave.

(a) State an assumption which leads to a simple relationship between microwave oven power and cooking times. What is this relationship?

Hint: if you have difficulty coming up with a relationship between power and time then you might like to think about the definition of the watt: one watt is one joule (a unit of energy) per second.

(b) Make an estimate of the range of validity of your model.

(c) In order to prevent confusion in the future, you decide to produce a table which lists the times for both microwaves for some items that you cook regularly. Use your relationship from part (a) to fill in the table, rounding your answers to the nearest half minute.

Food item	Cooking time (minutes)	
	500-watt oven	650-watt oven
One corn-on-the-cob		5
225 g oven chips	9	
Whole kipper		6
4×225 g cod steaks		12
2 kg beef	64	
2×100 g pork chops	6	

Question 6

The size of televisions is often quoted as the length of the diagonal of the screen. However, people often subconsciously judge the size of the screen by the screen area.

(a) How much bigger would someone perceive a 24-inch television to be compared with a 21-inch television? (Assume that both television screens are the traditional shape.)

(b) Generalize your answer of the previous part from a 24-inch television to an x-inch television. How much bigger is an x-inch television perceived to be compared with a 21-inch model?

Question 7

The volume of a sphere of radius r is $\frac{4}{3}\pi r^3$ and the surface area is $4\pi r^2$. Suppose that you are a manufacturer of spherical sweets and you already make sweets 1 cm in diameter. Some market research has indicated that there would be a market for a giant version of these sweets with a diameter of 3 cm. Both sizes of sweets consist of a fudge core with a very thin sugar coating.

(a) By what factor would you scale up the ingredients for the coating and the interior of one sweet?

(b) There are currently 600 small sweets to a packet. How many giant sweets would need to be put into a packet of about the same weight?

(c) Use your answers to the previous two parts of this question to estimate by how much the ingredients for the interiors and the coatings need to be scaled for one packet of giant sweets.

(d) The sweets are actually manufactured in batches of 100 packets. What would the scaling of the ingredients be for each batch of the giant sweets?

Question 8

A DIY enthusiast is trying to estimate how long it would take to dig a swimming pool in the garden. It takes 2 hours to dig a test hole which measures 1 m by 1 m by 0.5 m.

(a) Devise a formula for how long it will take to dig a hole which measures x m by x m by $x/2$ m.

(b) How long will it take to dig the whole swimming pool, which measures 5 m by 5 m by 2.5 m?

Question 9

One method that astronomers use to find the distance from the Earth to a star is to use a relationship between the apparent brightness and the distance from the Earth. The apparent brightness of a star viewed from the Earth is determined by the intensity of light reaching the Earth. Astronomers have instruments for measuring the intensity of light from stars.

Intensity of light is defined to be the amount of light energy hitting a unit area per second.

Now consider a star which emits a total light energy of E every second. Assume that the light from the star radiates equally in all directions. So when the light has travelled a distance r the light will be spread over the surface of an imaginary sphere as in Figure 4.

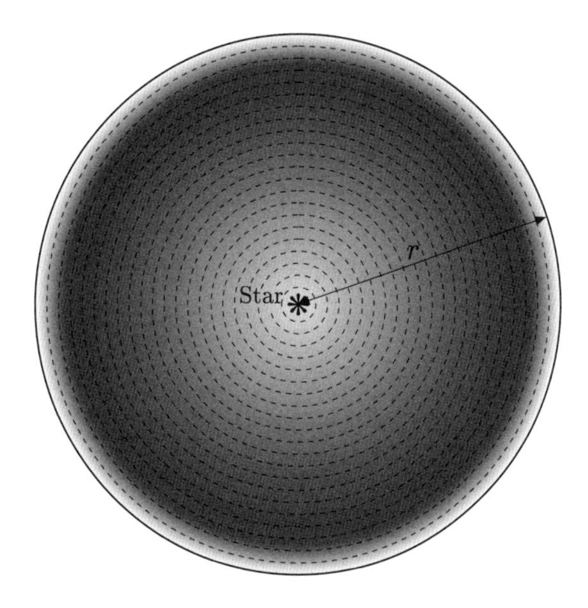

Figure 4

Let the intensity of the light at any point of unit area on the surface of this imaginary sphere be I.

(a) Calculate the total light energy per second meeting the surface of the imaginary spherc in terms of I and r. (The formula for the surface area of a sphere of radius r is $4\pi r^2$.)

(b) Since energy must be conserved, the energy you calculated in the previous part must equal E, the total light energy emitted by the star per second. Rearrange the formula to make I the subject.

Question 10

It is useful to be able to recognize the graphs of the common functions. Practise your skill now by trying to match a function with its graph. The six types of functions that you have to choose from are the following.

1 A sinusoidal function of the form $y = A\sin x$.

2 An exponentially decreasing function of the form $y = ab^x$ with $b < 1$.

3 An exponentially increasing function of the form $y = ab^x$ with $b > 1$.

4 A quadratic function of the form $y = a(x - k)^2 + l$.

5 A linear function of the form $y = mx + c$.

6 A inversely proportional function of the form $y = k/x$.

The graphs of these functions are shown in Figure 5.

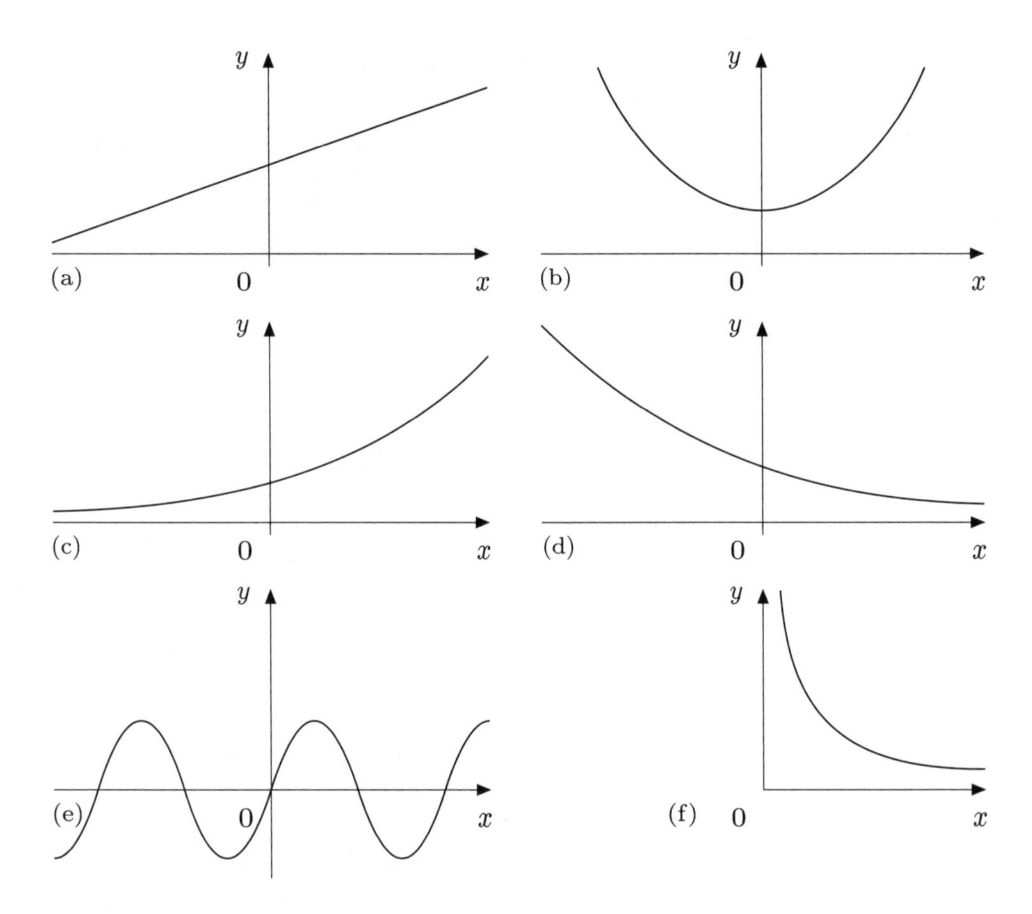

Figure 5

Write down your answer as a sequence of pairs: for example, if you think that the sinusoidal function is represented by graph (b) then write 1b.

Solutions

Unit 10

1

(a) Figure 6 shows an appropriate linear graph.

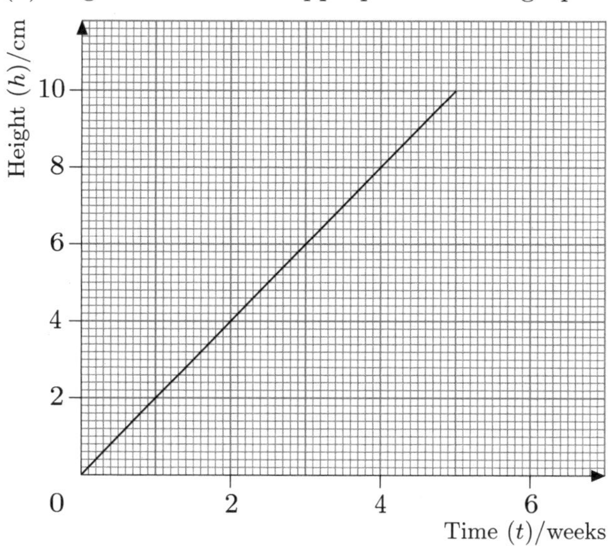

Figure 6

(b) The question you want to answer is: 'when does the pile of mail reach one centimetre from the top of the pigeon-hole—that is, when does $h = 9$ cm?'

(c) From the graph, $h = 9$ cm when $t = 4.5$ weeks. This means that the secretary must empty the pigeon-hole at least once every $4\frac{1}{2}$ weeks.

(d) The gradient of the graph = 2 cm per week, which is the rate at which the height of the pile of mail grows.

2

(a) Figure 7 shows an appropriate linear graph.

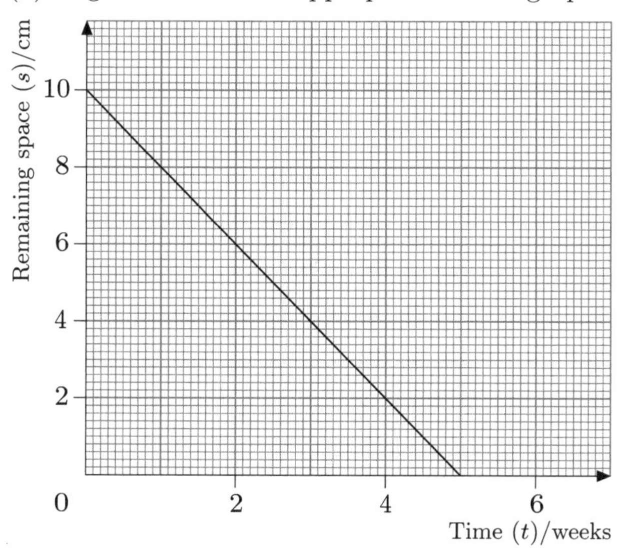

Figure 7

(b) The question you want to answer is: 'when is the space left in the pigeon-hole equal to one centimetre—that is, when is $s = 1$ cm?'

(c) From the graph, $s = 1$ cm when $t = 4.5$ weeks. This means that the secretary must empty the pigeon-hole at least once every $4\frac{1}{2}$ weeks.

(This is the same result as the previous question. This *must* happen because no aspect of the real problem changed; only our mathematical model of the problem has changed.)

(d) The gradient of the graph = $^{-}2$ cm per week. This gradient is negative, and it means that the amount of space left is *decreasing* at a rate of 2 cm per week.

3

(a) Let the temperature be T °C at time h hours after sunset.

(You may have chosen other letters. In particular you may have chosen t instead of h; but try to avoid confusion between T and t.)

(b) Assume the temperature drops at a constant rate of 0.5 °C per hour.

(c) The initial value is the value of T when $h = 0$. This is 5.

(d) $T = 5 - 0.5h$.

(e) The model is only valid between sunset ($h = 0$) and sunrise (probably around $h = 12$).

(f) The temperature has dropped to freezing point when $T = 0$. So you need to use the equation to find the value of h when $T = 0$.

(g) Substitute $T = 0$ into the equation:

$$0 = 5 - 0.5h$$
$$0.5h = 5$$
$$h = 10$$

The temperature will fall to 0 °C ten hours after sunset, that is at 4.30 am. The workers would, however, need to light fires *before* this. They would probably allow a good safety margin, and get up and light the fires between 3.30 and 4.00 am.

4

(a)

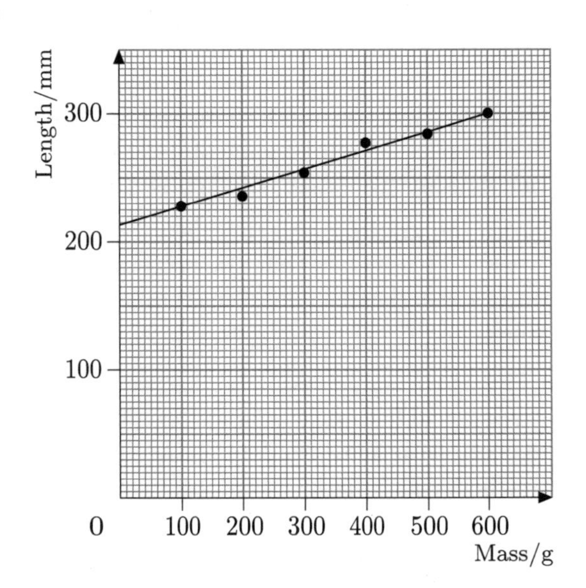

Figure 8

(b) (i) The intercept of the graph is at 213 mm (your value should be something similar, say plus or minus 5): this is the length of the elastic band before it was stretched.

(ii) The increase in length per additional 1 g mass is about 0.15 mm. To calculate this, choose two points on your 'best fit' line which are a reasonable distance apart and calculate the change in the length divided by the change in the mass: for example, if the points (100, 228) and (600, 301) are on your line, then the gradient is:

$$\frac{301-228}{600-100}=\frac{73}{500}=0.15$$

(to two decimal places)

(c) The equation of the 'best fit' line for the values found in part (b) is:

$$y = 0.15x + 213.$$

(You may have slightly different values.)

(d) The calculator produces a regression line with equation $y = ax + b$ with $a = .1525714286$, $b = 210.6$. This should confirm the results you found in part (b). The predicted lengths are
(i) 264 mm (ii) 333 mm (iii) 394 mm.

Of course, all of these values assume that the linear model for expansion is valid. This is probably true for the 350 g mass, since this is inside the range of the given data (interpolation). The other two masses, however, are outside the given data range (extrapolation), and so the model may not be valid for these values.

In fact, for large masses over a certain mass (known as the elastic limit), the linear model fails (for example, in this case the elastic band actually snapped when the 1200 g mass was attached).

5

(a) Define the variables as distance d miles from Edinburgh and time t hours after 10.30 am.

Assuming that the goods train is travelling at a constant speed of 50 mph starting at $d = 0$ and $t = 0$, its motion can be modelled by a straight line of gradient 50 passing through the origin $(0, 0)$. The model is valid until either the train reaches Newcastle or it is pulled off the main line.

Assuming that the holiday train is travelling at a constant speed of 90 mph, then its motion can also be modelled by a straight line. However, the holiday train sets off half an hour later, so it starts when $t = 0.5$ and $d = 0$. Its motion is therefore modelled by a straight line passing through the point $(0.5, 0)$ and with a gradient of 90. The model is valid until the train reaches Newcastle.

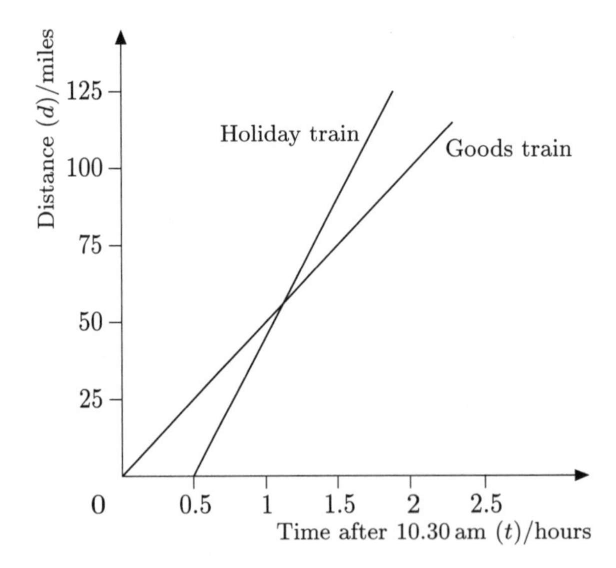

Figure 9

(b) The necessary equations for the calculator are $Y1 = 50X$ and $Y2 = 90X - 45$.

In order to find the equation of the graph representing the holiday train it is necessary to work out the gradient of the graph and where it would cross the vertical axis (the intercept). The gradient is the speed of the train, 90 mph and since when $t = 0.5$, $d = 0$, it follows that the intercept would be at $90 \times {}^{-}0.5 = {}^{-}45$.

The point of intersection of the two lines (see *Calculator Book*, Section 8.6) is at $X = 1.125$, $Y = 56.25$. So the model predicts that the trains would meet about 56 miles from Edinburgh, about 1.125 hours or 1 hour 7.5 minutes after 10.30 am, i.e at just before 11.40 am.

(c) No allowance was made for variation in speed or for the time needed to re-route the goods train. So, a good safety margin is needed. The goods train must be pulled off the main line well before 56 miles, perhaps at a suitable place about 50 miles from Edinburgh.

6 Let d be the distance from Folkestone measured in miles and let t be the time after 20.50 measured in minutes. (You could equally well choose to measure the time in minutes after 20.15.) Both ferry journeys can be modelled by linear models if you assume that each ferry travels at a constant speed. Each model will be valid for times less than or equal to the 110-minute journey time (and also for distances up to 30 miles).

For your ferry, $d = 0$ when $t = 0$ and $d = 30$ when $t = 110$. So your ferry's journey is represented by the straight line through the two points $(0, 0)$ and $(110, 30)$.

For the other ferry, the timetable says that the ferry was in Calais harbour at 20.15; so $d = 30$ when $t = {}^{-}35$. It is due to arrive at Folkestone at 22.05; so $d = 0$ when $t = 75$. So the other ferry's journey is represented by the straight line through the two points $({}^{-}35, 30)$ and $(75, 0)$.

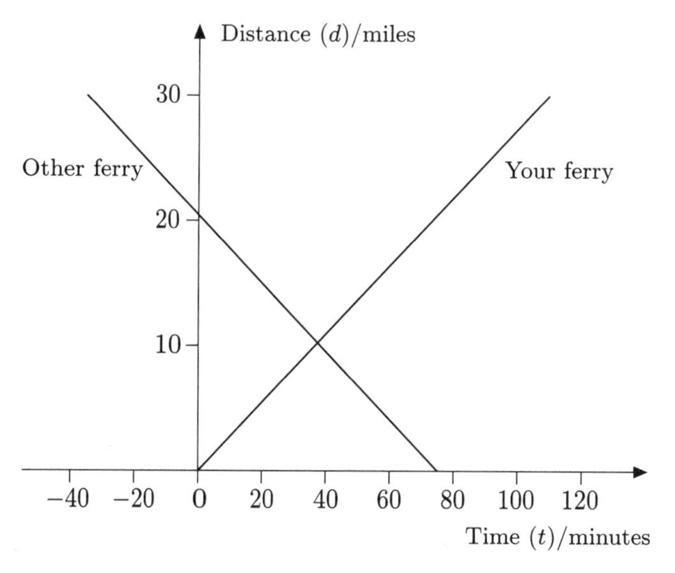

Figure 10

Both ferries are travelling at an average speed of $30 \div 110 = 3/11$ miles per minute but, since they are heading in opposite directions, the gradients of the two lines are $3/11$ and ${}^{-}(3/11)$. It is necessary to work out the value of the intercept for the graph representing the second ferry. Since $d = 30$ when $t = {}^{-}35$ the value of the intercept $(t = 0)$ is $30 - 35 \times (3/11)$.

The necessary equations can be entered on the calculator as $Y1 = (3/11)X$ and $Y2 = {}^{-}(3/11)X + 30 - 35 \times (3/11)$.

The point of intersection of the two lines gives the position and time where the two ferries pass. Using the calculator's intersect command (see *Calculator Book*, Section 8.6) gives the coordinates of this point as $X = 37.5$, $Y = 10.2$ (to 1 d.p). This means that it occurs 37.5 minutes into your journey. The model is not very accurate and so it might be safest to round down to the nearest 10 minutes. So you should look out for the other ferry from about 21.20 onwards.

7

(a)
$$y = 3 + x \qquad (1)$$
$$y = 3x + 7 \qquad (2)$$

Substitute for y in Equation (2) using Equation (1) and simplify:

$$\begin{aligned} 3 + x &= 3x + 7 \\ {}^{-}4 &= 2x \\ x &= {}^{-}2 \end{aligned}$$

Substitute this value into Equation (1) to get:

$$y = 3 + {}^{-}2 = 1$$

So the solution is: $x = {}^{-}2$, $y = 1$. Checking in Equation (2) gives: $1 = 3({}^{-}2) + 7$, which is correct.

(b)
$$y = 2 - 3x \qquad (3)$$
$$3y = {}^{-}4 - 4x \qquad (4)$$

Substitute for y in Equation (4) using Equation (3) and simplify:

$$\begin{aligned} 3(2 - 3x) &= {}^{-}4 - 4x \\ 6 - 9x &= {}^{-}4 - 4x \\ 6 + 4 &= {}^{-}4x + 9x \\ 10 &= 5x \\ x &= 2 \end{aligned}$$

Substitute this value into Equation (3) to get:

$$\begin{aligned} y &= 2 - 3(2) \\ &= 2 - 6 \\ &= {}^{-}4. \end{aligned}$$

So the solution is: $x = 2$, $y = {}^{-}4$.

Checking in Equation (4) gives:
$3(^-4) = {}^-4 - 4(2)$, which is correct.

(c)
$$x = y - 1 \quad (5)$$
$$3y = 2x - 2 \quad (6)$$

Substitute for x in Equation (6) using Equation (5) and simplify:

$$\begin{aligned} 3y &= 2(y-1) - 2 \\ &= 2y - 2 - 2 \\ y &= {}^-4. \end{aligned}$$

Substitute this value into Equation (5) to get:

$$x = {}^-4 - 1 = {}^-5.$$

So, the solution is $x = {}^-5$, $y = {}^-4$.
Checking in Equation (6) gives:
$3(^-4) = 2(^-5) - 2$, which is correct.

(d)
$$y = 3 - 2x \quad (7)$$
$$y = 2x - 1 \quad (8)$$

Substitute for y in Equation (8) using Equation (7) and simplify:

$$\begin{aligned} 3 - 2x &= 2x - 1 \\ 4 &= 4x \\ 1 &= x \end{aligned}$$

Substitute this into Equation (7) to get:

$$y = 3 - 2 \times 1 = 1$$

So the solution is $x = 1$ and $y = 1$.

Checking in Equation (8) gives
$1 = 2 \times 1 - 1$, which is correct.

(e)
$$y = 10 + 5x \quad (9)$$
$$2y = 3x + 1 \quad (10)$$

Substitute for y in Equation (10) using Equation (9) and simplify:

$$\begin{aligned} 2 \times (10 + 5x) &= 3x + 1 \\ 20 + 10x &= 3x + 1 \\ 7x &= {}^-19 \\ x &= -\frac{19}{7} \end{aligned}$$

Substitute this into Equation (9) to get:

$$y = 10 + 5 \times \left(-\frac{19}{7}\right) = -\frac{25}{7}$$

So the solution is $x = {}^-2.71$ and $y = {}^-3.57$ (rounded to two decimal places).

Checking in Equation (10) gives

$$2 \times \left(-\frac{25}{7}\right) = 3 \times \left(-\frac{19}{7}\right) + 1$$

which is correct.

(f)
$$2y + x = 7 \quad (11)$$
$$3y + 2x = 4 \quad (12)$$

Rearrange Equation (11) to make y the subject:

$$\begin{aligned} 2y &= 7 - x \\ y &= 3.5 - 0.5x \end{aligned}$$

Substitute this into Equation (12) and simplify:

$$\begin{aligned} 3 \times (3.5 - 0.5x) + 2x &= 4 \\ 10.5 - 1.5x + 2x &= 4 \\ 0.5x &= {}^-6.5 \\ x &= {}^-13 \end{aligned}$$

Substitute this into Equation (11) and solve for y:

$$\begin{aligned} 2y + (^-13) &= 7 \\ 2y &= 20 \\ y &= 10 \end{aligned}$$

So the solution is $x = {}^-13$ and $y = 10$.

Checking in Equation (12) gives
$3 \times 10 + 2 \times (^-13) = 4$, which is correct.

8

(a) Looking at the models for supply and demand separately gives the following.

Supply

The initial value (intercept) is $^-5$.
The gradient is 1.
The equation is:

$$y = x - 5 \quad (13)$$

Demand

The initial value (intercept) is 40 (to the nearest number).
The gradient is $^-0.7$.
The equation is:

$$y = 40 - 0.7x \quad (14)$$

(b) The equilibrium price is where supply and demand are equal—that is, at the values of x and y which satisfy the equation for supply (13) and demand (14) simultaneously. So solve the simultaneous Equations (13) and (14). Substitute for y in Equation (14) using Equation (13) to give:

$$\begin{aligned} x - 5 &= 40 - 0.7x \\ 1.7x &= 45 \\ x &= \frac{45}{1.7} \\ x &= 26.5 \quad \text{(to one decimal place)} \end{aligned}$$

So the equilibrium price is around 26 to 27p.

(c) Equation (13) gives $y = 21.5$ (to one decimal place). Checking in Equation (14) also gives $y = 21.5$ (to one decimal place). So about 21.5 million kg of soft fruit will be sold.

(d) It has been assumed that future supply and demand will be the same as it was in the past. Linear models have also been assumed. These assumptions do not take into account any change in such factors as inflation, bad weather, seasons. Hence the predictions of the equilibrium price and of the corresponding amount of soft fruit sold are only approximate.

9

(a)

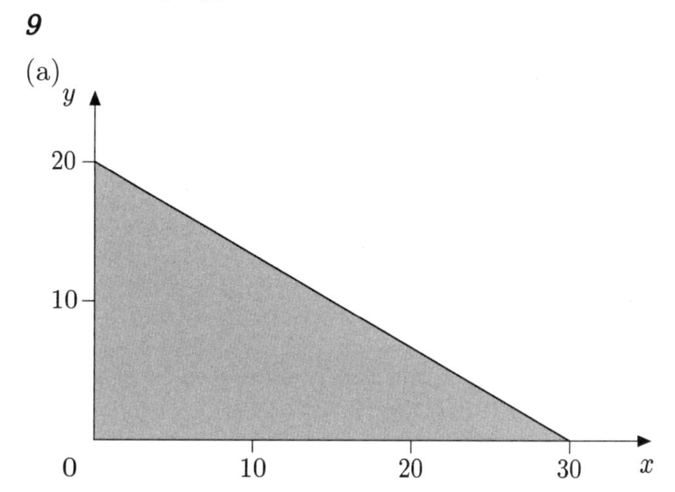

Figure 11

(b)

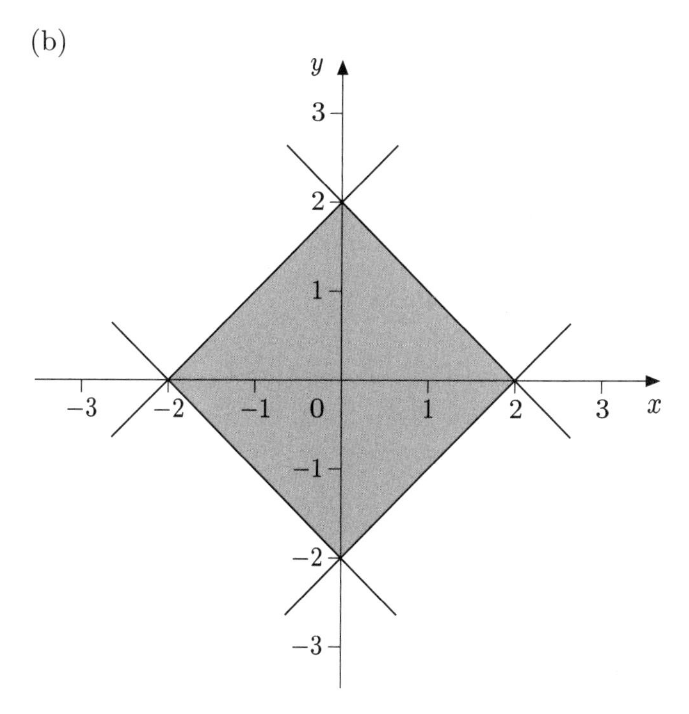

Figure 12

(c)

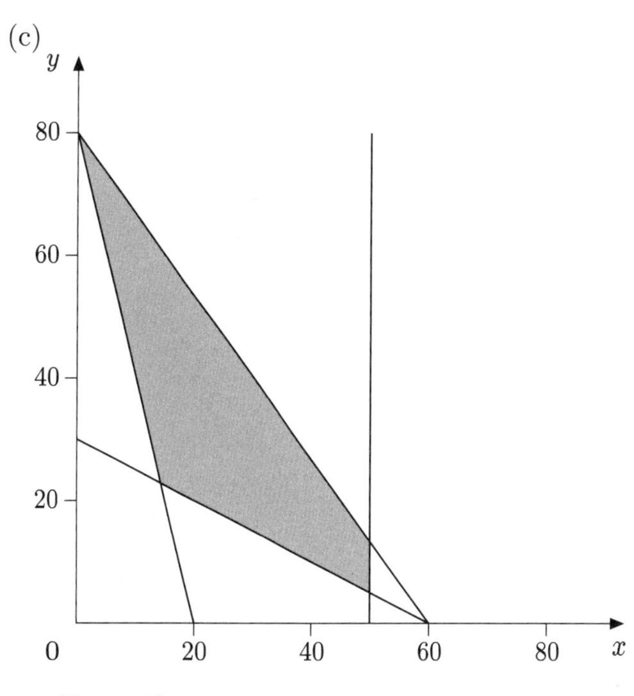

Figure 13

10

(a) The feasible region is sketched in Figure 14.

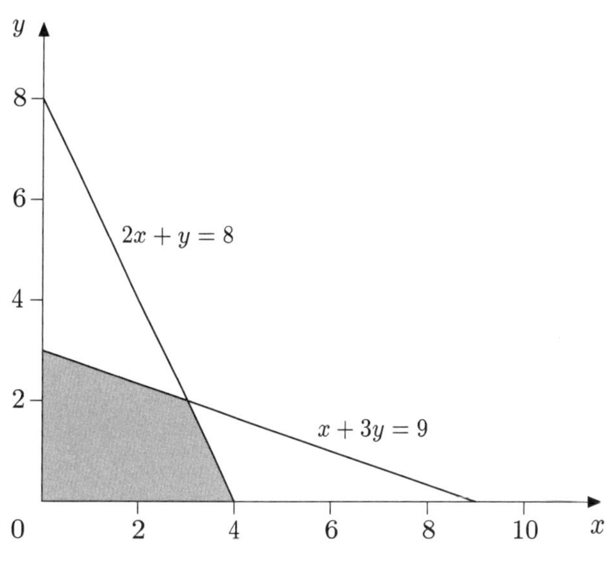

Figure 14

From the diagram the vertices can be read off as $(0, 0)$, $(0, 3)$, $(3, 2)$, $(4, 0)$.

(b) (i) By drawing in a few P-lines it is clear that the vertex $(3, 2)$ is the point that maximizes $P = x + y$ and the value of P is then $3 + 2 = 5$.

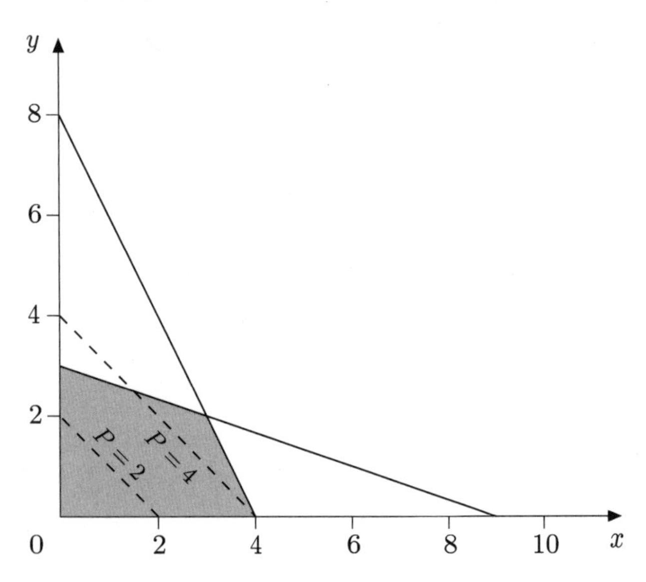

Figure 15

(ii) By drawing in a few P-lines it is clear that the vertex $(4, 0)$ maximizes $P = 3x + y$ and the value of P is then $3 \times 4 + 0 = 12$.

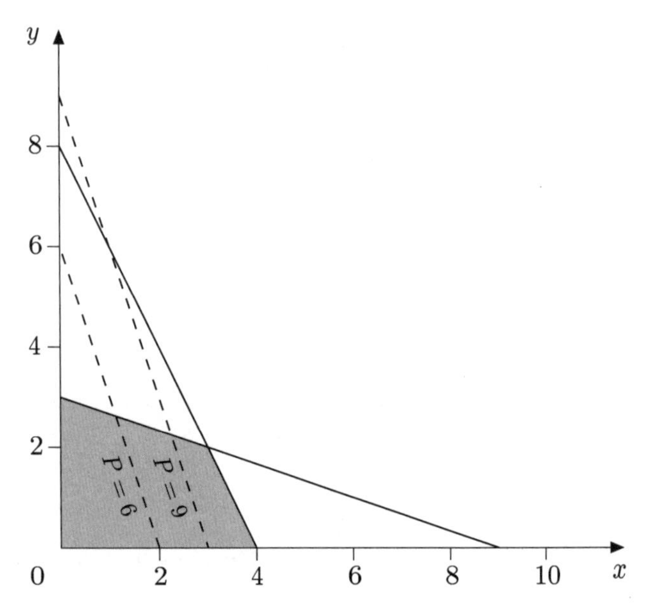

Figure 16

11

(a) The feasible region is sketched in Figure 17.

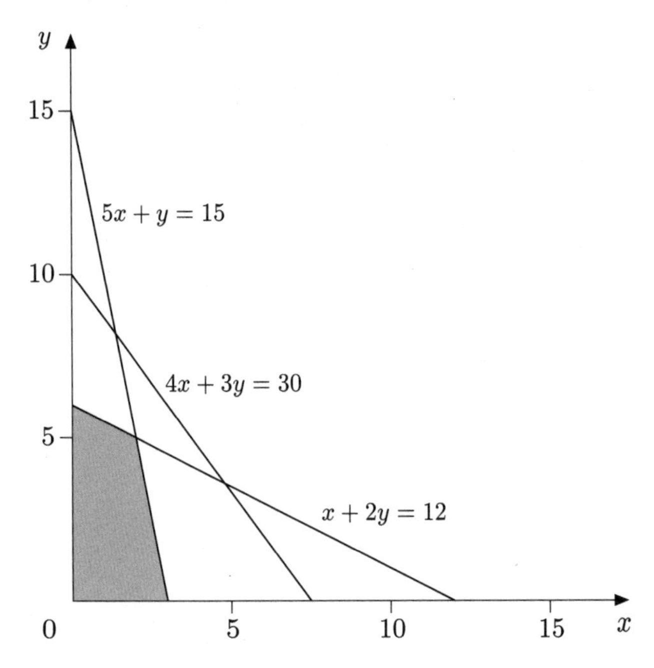

Figure 17

From the diagram the vertices can be read off as $(0, 0)$, $(0, 6)$, $(2, 5)$, $(3, 0)$.

(b) (i) By drawing in a few P-lines it is clear that the vertex $(2, 5)$ maximizes $P = x + y$ and the value of P is then $2 + 5 = 7$.

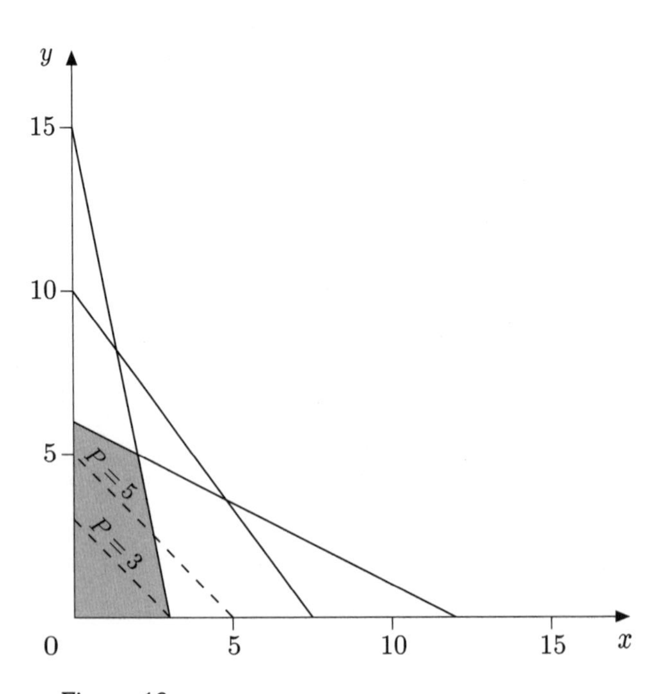

Figure 18

(ii) By drawing in a few P-lines it is clear that the vertex $(2, 5)$ maximizes $P = 7x + 2y$ and the value of P is then $7 \times 2 + 2 \times 5 = 24$.

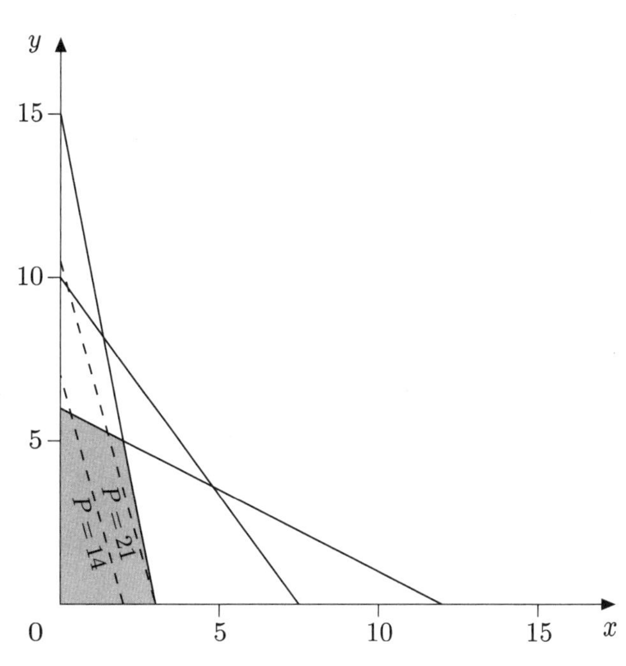

Figure 19

12

(a) Start by defining variables for the problem. Let x be the number of fruit cakes made and let y be the number of iced cakes made.

The information given in the question can be summarized by the following table.

	Fruit cakes x	Iced cakes y	Total available
Preparation time (hours)	1	2	18
Oven time (hours)	1.5	1	14
Price (£)	0.80	1.00	

The restriction that the number of preparation hours is limited to 18 translates to the following inequality:

$$x + 2y \leq 18$$

Similarly, the restriction that the number of oven hours is limited to 14 translates to the following inequality:

$$1.5x + y \leq 14$$

Further constraints are implicit in the question. Obviously the number of each type of cake must be positive, so that $x \geq 0$ and $y \geq 0$. Also only whole numbers of cakes can be made.

The function to be maximized is the income:

$$P = 0.8x + y$$

The feasible region and two P-lines are shown in Figure 20.

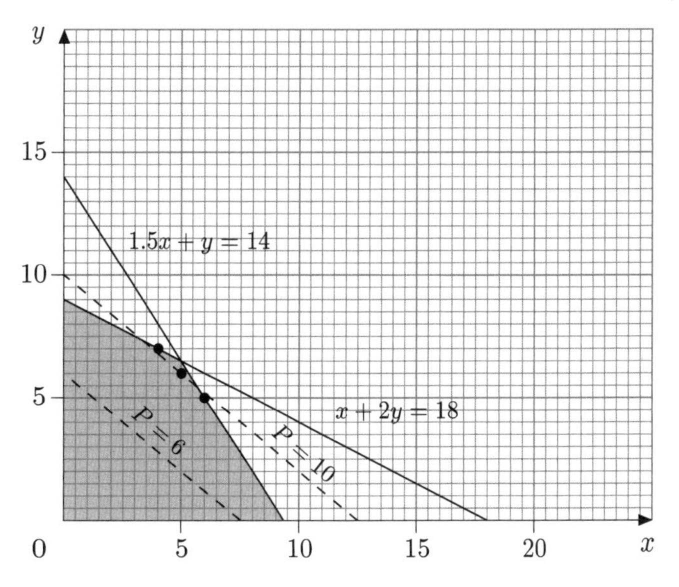

Figure 20

So, clearly, the maximum value occurs at the vertex where the lines $x + 2y = 18$ and $1.5x + y = 14$ intersect. By solving this pair of simultaneous equations or by examining the graph, the coordinates of this vertex are found to be $(5, 6.5)$; and the y-coordinate is non-integer.

To find the optimal integer-valued solution, examine the integer-valued points near this vertex. Bearing in mind the $P = 10$ line, the candidates are the points $(4, 7)$, $(5, 6)$ and $(6, 5)$ marked on Figure 20. From the figure, it looks as if it is probably $(4, 7)$. Check:

at $(4, 7) : P = 0.8 \times 4 + 7 = 10.2$;
at $(5, 6) : P = 0.8 \times 5 + 6 = 10$;
at $(6, 5) : P = 0.8 \times 6 + 5 = 9.8$.

The gives the optimal solution that 4 fruit cakes and 7 iced cakes should be made to yield an income of £10.20.

(b) Changing the price of the iced cakes changes the function which is to be maximized to:

$$P = 0.8x + 1.5y$$

It does not change the feasible region. Two P-lines corresponding to the new P-function are shown in Figure 21.

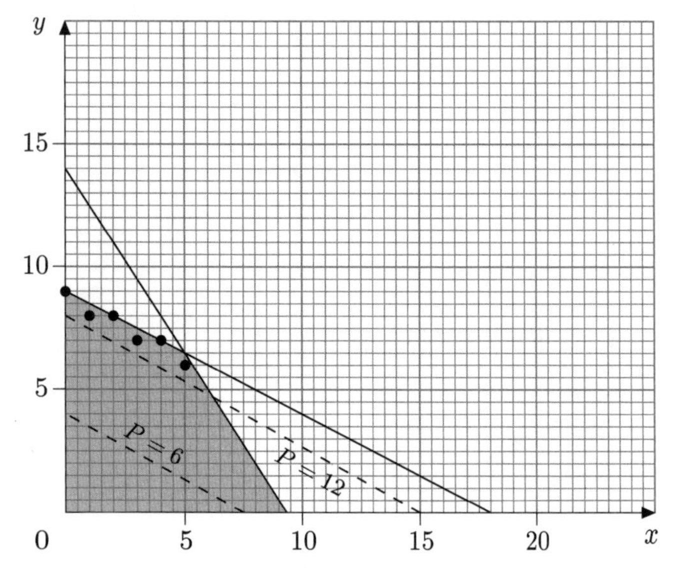

Figure 21

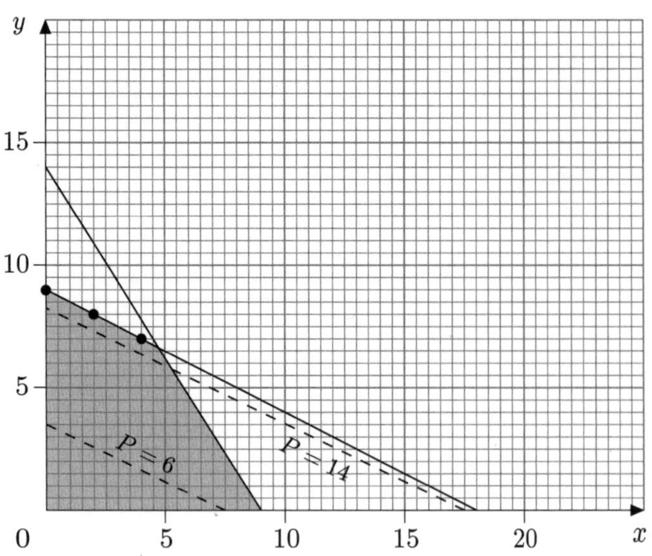

Figure 22

Again, the maximum value occurs at the vertex $(5, 6.5)$, which is non-integer. The candidates for the best integer-valued point are $(0, 9)$, $(1, 8)$, $(2, 8)$, $(3, 7)$, $(4, 7)$ and $(5, 6)$. The points are marked on Figure 21. Clearly $(2, 8)$ is better than $(1, 8)$ and $(4, 7)$ is better than $(3, 7)$. Checking the other four points:

$$\text{at } (0, 9): P = 0.8 \times 0 + 1.5 \times 9 = 13.5;$$
$$\text{at } (2, 8): P = 0.8 \times 2 + 1.5 \times 8 = 13.6;$$
$$\text{at } (4, 7): P = 0.8 \times 4 + 1.5 \times 7 = 13.7;$$
$$\text{at } (5, 6): P = 0.8 \times 5 + 1.5 \times 6 = 13.$$

So, to maximize your income, you should make 4 fruit cakes and 7 iced cakes to yield £13.70. So, increasing the price of iced cakes to £1.50 does not change the optimum numbers of cakes that you should bake.

(c) The new cost function is:

$$P = 0.8x + 1.7y$$

Figure 22 shows two P-lines corresponding to this new cost function.

This time it looks like the maximum value occurs at a different vertex, at the point $(0, 9)$ where $x + 2y = 10$ cuts the y-axis. This point is integer-valued. But, to check, because the P-lines are close to being parallel to $x + 2y = 10$, it is worth evaluating P at the points $(0, 9)$, $(2, 8)$ and $(4, 7)$ shown on the figure.

$$\text{at } (0, 9): P = 0.8 \times 0 + 1.7 \times 9 = 15.3;$$
$$\text{at } (2, 8): P = 0.8 \times 2 + 1.7 \times 8 = 15.2;$$
$$\text{at } (4, 7): P = 0.8 \times 4 + 1.7 \times 7 = 15.1.$$

The change in price to £1.70 for an iced cake has now made a difference to the optimum number of cakes. It is now more advantageous, providing an income of £15.30, to make 9 iced cakes and no fruit cakes, as long as this is acceptable to the club.

13

(a) The restrictions given in the question translate to the following constraints:

$$\begin{aligned} 8x + 16y &\geq 75 \quad \text{(protein)} \\ 46x + 6y &\geq 250 \quad \text{(carbohydrate)} \\ 2x + 8y &\geq 70 \quad \text{(fat)} \end{aligned}$$

Also implicit in the question are the restrictions that the amount of bread and cheese in the diet are positive, so that $x \geq 0$ and $y \geq 0$. The daily calorie intake, which must be minimized, is given by:

$$P = 235x + 166y$$

Figure 23 shows the feasible region.

(b) The line representing the minimum protein requirement does not intersect the feasible region. The conclusion that can be drawn from this is that the protein requirement is automatically satisfied if the other requirements are met.

(c) Two P-lines for decreasing values of P are shown in Figure 24. From the figure, the point that minimizes the calorie intake is at the intersection of the lines representing the minimum carbohydrate and fat requirements—that is, at approximately $(4.4, 7.6)$. Interpreting this solution in terms of the original problem gives the optimum diet of eating 440 g of bread and 760 g of cheese per day.

[You may have slightly different values from your diagram.]

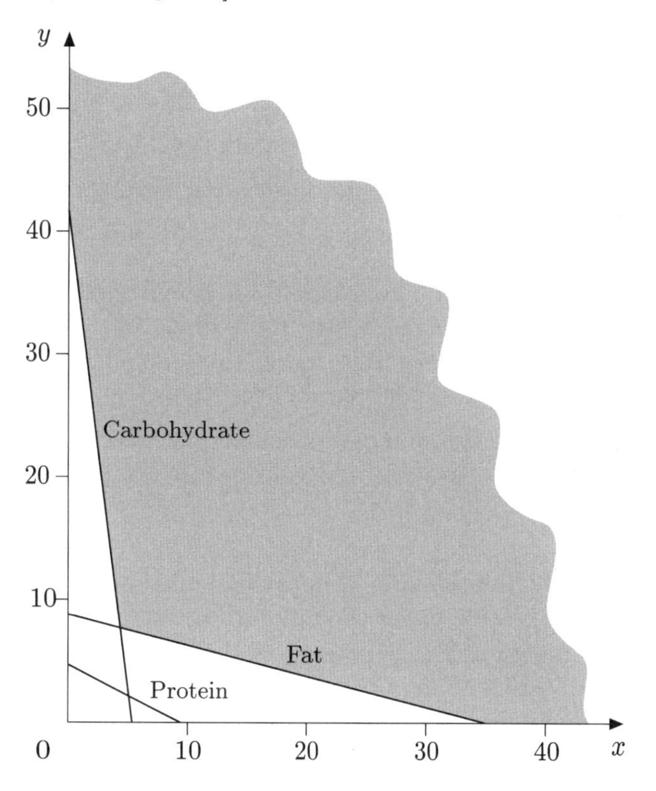

Figure 23

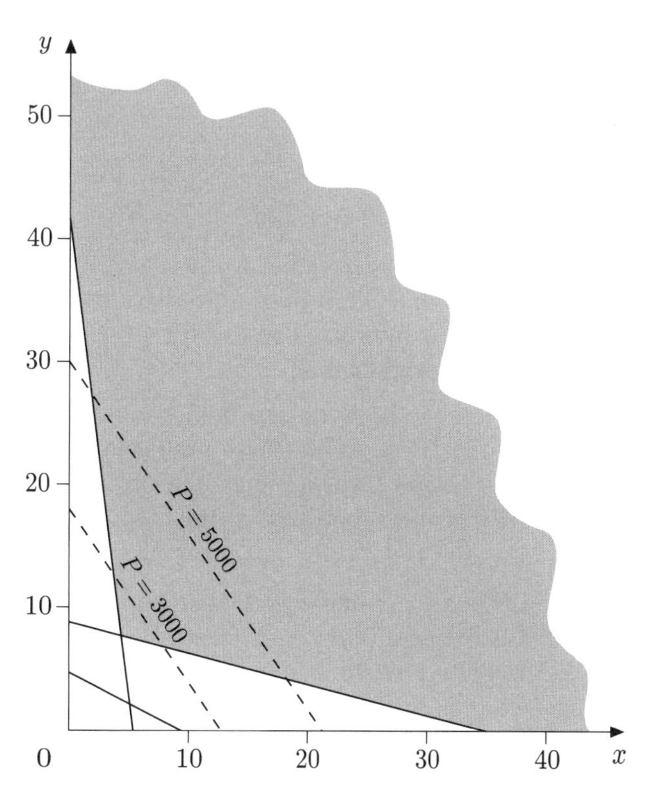

Figure 24

(d) The coordinates of the optimal solution can be worked out by solving simultaneously the equations of the lines corresponding to the minimum carbohydrate requirement and the minimum fat requirement.

$$46x + 6y = 250 \qquad (15)$$
$$2x + 8y = 70 \qquad (16)$$

Rearranging Equation (16) gives

$$x = 35 - 4y \qquad (17)$$

which can be substituted into Equation (15):

$$46\,(35 - 4y) + 6y = 250$$
$$1610 - 184y + 6y = 250$$
$$^{-}178y = {}^{-}1360$$
$$y = 7.64 \quad \text{(to two d.p.)}$$

From Equation (17),

$$x = 35 - 4 \times 7.64 = 4.44 \quad \text{(to 2 d.p.)}$$

Interpreting this solution in terms of the original problem gives the optimum diet of eating 444 g of bread and 764 g of cheese per day.

(e) The daily calorie intake for this diet is $235 \times 4.44 + 166 \times 7.64 = 2310$ calories per day (rounded to the nearest ten).

Unit 11

1

(a) To move a parabola k units along the x-axis, substitute $(x - k)$ for every occurrence of x in the equation. Similarly, to move a parabola l units along the y-axis, substitute $(y - l)$ for every occurrence of y in the equation.

(i) To move the parabola with general equation $y = ax^2$ by 4 units along the y-axis to $(0, 4)$, substitute $(y - 4)$ for y in the equation:

$$(y - 4) = ax^2$$
$$y = ax^2 + 4$$

(ii) To move the parabola with general equation $y = ax^2$ by 4 units along the x-axis to $(4, 0)$, substitute $(x - 4)$ for x in the equation:

$$y = a(x - 4)^2$$

(iii) To move the parabola with general equation $y = ax^2$ so that its vertex is $(1, {}^-2)$, substitute $(x - 1)$ for x and $(y - ({}^-2)) = (y + 2)$ for y in the equation:

$$(y + 2) = a(x - 1)^2$$
$$y = a(x - 1)^2 - 2$$

(b) Moving the general equation of a parabola with the vertex at the origin $y = ax^2$ to one with its vertex at $(6, {}^-4)$ gives:

$$(y - ({}^-4)) = a(x - 6)^2$$
$$(y + 4) = a\left(x^2 - 12x + 36\right)$$
$$(y + 4) = ax^2 - 12ax + 36a$$
$$y = ax^2 - 12ax + 36a - 4$$

Now to obtain the equations of two parabolas with vertex at $(6, {}^-4)$ you can pick any two non-zero values for a.

$a = 1$ gives $y = x^2 - 12x + 32$
$a = 2$ gives $y = 2x^2 - 24x + 68$

(You may have chosen other values for a.)

You could use your calculator to check your answers to this question, by drawing the graphs of your chosen parabolas.

2 Begin with the general parabola with vertex at the origin $(y = ax^2)$ and move it to find the equation of the general parabola with vertex at $({}^-1, 2)$:

$$(y - 2) = a(x - ({}^-1))^2$$
$$y = a(x + 1)^2 + 2$$

Now substitute in $x = 1$ and $y = 14$ to get an equation for the unknown a:

$$14 = a \times 2^2 + 2 = 4a + 2$$

So $4a = 12$, and this gives $a = 3$. So the equation of the desired parabola is:

$$y = 3(x + 1)^2 + 2$$

This multiplies out to:

$$y = 3x^2 + 6x + 5$$

3

(a) This quadratic equation can be rearranged into the form:

$$(y - 1) = (x + 1)^2$$

This equation can be obtained by moving the parabola $y = x^2$ by ${}^-1$ along the x-axis and 1 along the y-axis. So the vertex $(0, 0)$ of the parabola $y = x^2$ has moved by ${}^-1$ along the x-axis and 1 along the y-axis to the point $({}^-1, 1)$. So the vertex of the given parabola is $({}^-1, 1)$.

(b) Rearrange the equation into the form:

$$(y - 1) = x^2$$

This is the equation obtained by moving the parabola $y = x^2$ one unit along the y-axis. So the vertex of the parabola is $(0, 1)$.

(c) Rearrange the equation into the form:

$$(y - 1) = (x - 2)^2$$

So the vertex of the parabola is $(2, 1)$.

(d) Rearrange the equation into the form:

$$(y - 2) = 2(x - 3)^2$$

This equation is obtained by moving the parabola $y = 2x^2$ three units along the x-axis and two units along the y-axis. So the vertex of the parabola is $(3, 2)$.

In all four cases you can check your answers by plotting the parabolas on your calculator.

4

(a) On the calculator a linear fit looks quite reasonable given the amount of random variation that is present in the data. In the absence of any other information, a linear model is a reasonable choice.

(b) The regression line is

$$y = 1080x - 323$$

(with parameters rounded to the nearest whole number). This gives a prediction for the weight at month 15 of 15900 g (to the nearest hundred grams).

(c) The best fit parabola is

$$y = 43x^2 + 564x + 796$$

(with parameters rounded to the nearest whole number). This gives a prediction for the weight in month 15 of 18900 g (to the nearest hundred grams).

(d) The difference between the predictions is quite large. Comparing these predictions with the actual measurement in month 15 would be a very good test of the surface-area theory of coral growth.

Even though the linear fit looks good and has a high correlation coefficient ($r = 0.987$), the predicted weight could be incorrect if the surface-area growth theory is a good model.

This shows the dangers inherent in fitting data without a theoretical model of the underlying behaviour.

5

(a) The height of the cliff is the position of the stone initially (at time $t = 0$). Reading this off the graph gives the answer 100 m.

(b) Initially the graph is horizontal so this must represent the time between starting the watch and throwing the stone. The length of this initial line represents 5 seconds.

(c) The cliff is 100 m high, so trace from 100 m on the height axis across to the curve and then down to the time axis. This means that at time $t = 10$ seconds the stone returns to the height of the cliff edge.

(d) The graph is at the highest point approximately 7.5 seconds after the watch was started and reached a height of about 135 m, above sea level, 35 m above the cliff top.

(e) Collecting the data from the previous parts of the question, you want the equation of the parabola through the points $(5, 100)$, $(10, 100)$ and $(7.5, 135)$. The calculator gives the quadratic regression curve passing through these points to be:

$$y = {}^-5.6x^2 + 84x - 180.$$

(f) One way to calculate the time is to plot the graph of $y = 21$ together with the quadratic regression curve and to find where the two graphs intersect. This gives a value of $x = 12.01$ (to 2 d.p.), indicating that the stone hit the ground about 12 seconds after the watch was started.

(g) You can calculate the gradient of the quadratic regression curve using the nDeriv command. (see *Calculator Book*, Section 11.2). This gives a value of $^-50.5$, which is interpreted as a speed of 50.5 ms^{-1} vertically downwards.

6

(a) To use your calculator, enter in its lists the coordinates of three points on the parabola three known points are where the graph cuts the axes: $(^-1000, 0)$, $(0, 0.015)$, $(1000, 0)$. Quadratic regression then gives the following best fit curve:

$$y = {}^-1.5 \times 10^{-8}x^2 + 0.015$$

Alternatively you can calculate the equation algebraically. The equation of a parabola with axis parallel to the y-axis and having vertex at (k, l) is:

$$y = a(x - k)^2 + l$$

The vertex of the given parabola is at $(0, 0.015)$, so the equation is of the form:

$$y = ax^2 + 0.015$$

Then the coordinates of one of the other known points (for example, $(1000, 0)$) are substituted into this equation to find a:

$$0 = a \times 1000^2 + 0.015$$

So $a = {}^-1.5 \times 10^{-8}$ and the parabola has the equation:

$$y = {}^-1.5 \times 10^{-8}x^2 + 0.015$$

(b) Calculating the population density when $x = 200$ gives:

$$y = {}^-1.5 \times 10^{-8} \times 200^2 + 0.015 = 0.0144$$

So the predicted population density 200 m from the centre of the town is 0.0144 people per square metre.

7

(a) Figure 25 shows how the population density is likely to vary with distance from the main road. The curve could be a parabola. It has been assumed that the population density is zero outside the village boundary.

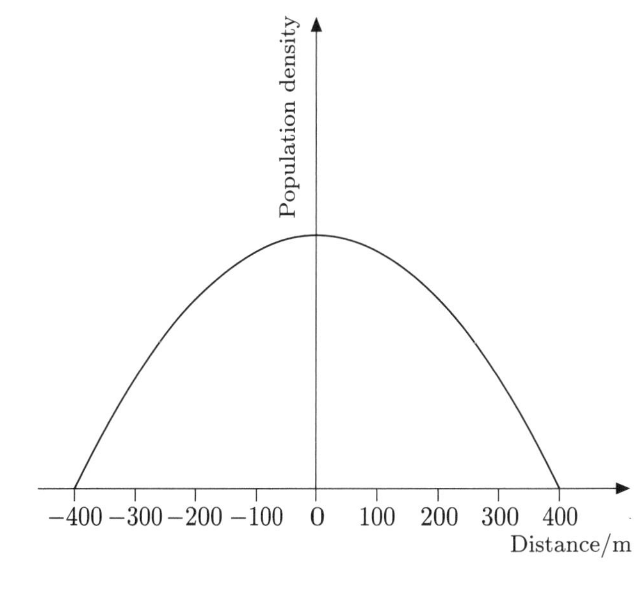

Figure 25

(b) The following assumptions are needed.

◇ The population density is zero outside the village (as stated above).

◇ The population density varies quadratically with the distance from the main road, and is symmetrical about the main road.

The variables will be the population density, y people m^{-2}, and the distance from the main road, x m. (Distances to the east will be positive, to the west negative.) The model will be valid only for $^{-}400 \leq x \leq 400$.

Using the calculator and quadratic regression, with the data points $(^{-}400, 0)$, $(0, 0.01)$ and $(400, 0)$, gives the equation:

$$y = {}^{-}6.25 \times 10^{-8}x^2 + 0.01$$

Alternatively, algebraically the equation relating y and x will be of the form:

$$y = a(x - k)^2 + l$$

As the peak population density is along the road and is 0.01 people m^{-2}, the vertex of the parabola is at $(0, 0.01)$ and so $k = 0$ and $l = 0.01$. The equation therefore becomes:

$$y = ax^2 + 0.01$$

The population is 0 at a distance of 400 m from the road, so $y = 0$ at $x = 400$. Using this gives:

$$0 = a(400)^2 + 0.01$$

Therefore:

$$a = \frac{^{-}0.01}{400^2} = {}^{-}6.25 \times 10^{-8}$$

Hence the equation is:

$$y = {}^{-}6.25 \times 10^{-8}x^2 + 0.01$$

This equation can be used to predict y at $x = 200$:

$$y = {}^{-}6.25 \times 10^{-8} \times 200^2 + 0.01$$
$$= 0.0075$$

So the population density is about 0.0075 people m^{-2} at 200 m from the road.

8

(a) (i) $x^2 = 36$

Taking the square root of both sides gives:

$$x = \pm\sqrt{(36)} = \pm 6.$$

(ii) $x^2 = 30$

Taking the square root of both sides gives:

$$x = \pm\sqrt{(30)} = \pm 5.48 \quad \text{(to 2 d.p.)}.$$

(iii) $5x^2 = 24$

Dividing both sides by 5 gives:

$$x^2 = 4.8$$

Taking the square root of both sides gives:

$$x = \pm\sqrt{(4.8)} = \pm 2.19 \quad \text{(to 2 d.p.)}.$$

(iv) $(x + 3)^2 = 18$

$$x + 3 = \pm\sqrt{(18)}$$
$$= \pm 4.24 \text{ (to 2 d.p.)}$$

$$x = 4.24 - 3 \text{ or } {}^{-}4.24 - 3$$
$$= 1.24 \text{ or } {}^{-}7.24 \quad \text{(to 2 d.p.)}$$

(b) (i) $4x^2 - 6x = 0$

Taking out the common factor of $2x$:

$$2x(2x - 3) = 0$$

There are two possible solutions:

$2x = 0$ or $2x - 3 = 0$
$2x = 3$
$x = 0$ or $x = 1.5$

(ii) $x^2 + 3x - 18 = 0$

This can be rewritten in its factorized form as:

$$(x - 3)(x + 6) = 0$$

There are two possible solutions:

$x - 3 = 0$ or $x + 6 = 0$
$x = 3$ or $x = {}^{-}6$

(iii) $2a^2 - 7a + 6 = 0$

This can be rewritten in its factorized form as:

$$(2a - 3)(a - 2) = 0$$

There are two possible solutions:

$2a - 3 = 0$ or $a - 2 = 0$
$2a = 3$
$a = 1.5$ or $a = 2$

(c) (i) $x^2 - 4x - 3 = 0$

Here, $a = 1$, $b = {}^{-}4$, $c = {}^{-}3$.

$$x = \frac{^{-}b \pm \sqrt{b^2 - 4ac}}{2a}$$
$$= \frac{^{-}(^{-}4) \pm \sqrt{(^{-}4)^2 - 4 \times 1 \times (^{-}3)}}{2 \times 1}$$
$$= \frac{4 \pm \sqrt{28}}{2}$$
$$= 4.65 \text{ or } {}^{-}0.65 \text{ (to two decimal places)}$$

(ii) $2x^2 - 7x + 4 = 0$

Here, $a = 2$, $b = {}^{-}7$, $c = 4$.

$$x = \frac{^{-}(^{-}7) \pm \sqrt{(^{-}7)^2 - 4 \times 2 \times 4}}{2 \times 2}$$
$$= \frac{7 \pm \sqrt{17}}{4}$$
$$= 2.78 \text{ or } 0.72 \text{ (to two decimal places)}$$

(iii) $3x^2 - 4x + 5 = 0$

Here, $a = 3$, $b = {}^-4$, $c = 5$.

$$x = \frac{^-(^-4) \pm \sqrt{(^-4)^2 - 4 \times 3 \times 5}}{2 \times 3}$$
$$= \frac{4 \pm \sqrt{^-44}}{6}$$

There are no real-number solutions because this expression includes the square root of a negative number.

Unit 12

1

(a) The population increases by tripling so the first four members of the sequence are 15, 45, 135, 405. The formula for the nth generation can most easily seen by writing down the first few terms in a way that makes the pattern evident.

Generation	Population
1	5×3
2	5×3^2
3	5×3^3
4	5×3^4
⋮	⋮
n	5×3^n

So the population in the 20th generation is $5 \times 3^{20} = 1.74 \times 10^{10}$ (to 3 s.f.).

(b) Adding 1% is equivalent to multiplying by 1.01. So the mass of the fungus (rounded to one decimal place) after the first four days is as shown below.

Generation	Mass (g)
1	$23.1 \times 1.01 = 23.3$
2	$23.1 \times 1.01^2 = 23.6$
3	$23.1 \times 1.01^3 = 23.8$
4	$23.1 \times 1.01^4 = 24.0$

This shows the pattern for the general formula. The mass of the fungus after the nth day is 23.1×1.01^n grams. This means that after the 20th day the mass of the fungus is 28.2 grams.

(c) Subtracting 10% is equivalent to multiplying by 0.9. So the balance after the first four months is shown below.

Generation	Balance (£)
1	$1000 \times 0.9 = 900$
2	$1000 \times 0.9^2 = 810$
3	$1000 \times 0.9^3 = 729$
4	$1000 \times 0.9^4 = 656.1$

This shows that the balance after n months is £1000×0.9^n, and so the balance after 20 months is £1000×0.9^{20} = £121.58 (to the nearest penny).

2 The general formula is $S(n) = a\left(\dfrac{b^n - 1}{b - 1}\right)$.

In this case $a = 1$ and $b = 1.01$ so the formula for the sum of the first n payments is

$$S(n) = 1\left(\frac{1.01^n - 1}{1.01 - 1}\right) = (1.01^n - 1) \div 0.01$$
$$= (1.01^n - 1) \times 100 = 100(1.01^n - 1)$$

So the total paid out in a 31-day month is $100(1.01^{31} - 1) = 36.13$ euros (to 2 d.p).

3

(a) The graphical method produces a value of $x = 1.301$ to three decimal places.

(b) 0.301029957, 1.30102996, 2.30102996.

(c) Remember that the process of finding the logarithm is 'undoing' the process of exponentiation. From parts (a) and (b) you know that

$$10^{0.301029957} = 2, \quad 10^{1.30102996} = 20,$$

and

$$10^{2.30102996} = 200.$$

It follows from the emerging pattern that $10^{3.30102996} = 2000$.

Alternatively, you could say that

$$10^{3.30102996} = 10^{(2.30102996+1)}$$
$$= 10^{2.30102996} \times 10^1$$
$$= 200 \times 10$$
$$= 2000.$$

4

(a) To find the population doubling time you need to solve the equation:

$$2 \times 209 = 209 \times 3^x.$$

This simplifies to:

$$2 = 3^x.$$

Taking logarithms of both sides:

$$\log_{10} 2 = \log_{10}(3^x)$$
$$\log_{10} 2 = x \times \log_{10} 3$$
$$\frac{\log_{10} 2}{\log_{10} 3} = x$$
$$x = 0.63 \quad \text{(to two decimal places)}$$

So the population doubling time is 0.63 time units.

(b) Using the formula for doubling time, d, with growth factor, b:

$$d = \frac{\log_{10} 2}{\log_{10} b}$$
$$= \frac{\log_{10} 2}{\log_{10} 1.2}$$
$$d = 3.80 \quad \text{(to two decimal places)}$$

So the population doubling time is 3.8 time units.

(c) Since the population is decreasing, the population will never double!

But if you did not realize this and applied the formula

$$d = \frac{\log_{10} 2}{\log_{10} b}$$
$$= \frac{\log_{10} 2}{\log_{10} 0.9}$$
$$d = {}^{-}6.58 \quad \text{(to two decimal places)}$$

The answer is negative, so this means that for the population to double you have to go back 6.58 time units—that is, 6.58 time units ago the population was twice what it is now. (So the half-life of the population is 6.58 time units.)

5

(a) The exponential model is expressed by an equation of the form:

$$y = a \times b^x$$

When $x = 0$ in the above equation, $y = a$. The initial population in this case is $y = 76$, so $a = 76$. Substituting this into the above equation gives:

$$y = 76 \times b^x$$

The population doubling time is 2 minutes, so $y = 2 \times 76$ when $x = 2$. Substituting these values into the equation gives:

$$2 \times 76 = 76 \times b^2$$

So $b^2 = 2$ and $b = \sqrt{2} \simeq 1.414$, and the exponential model equation is:

$$y = 76 \times (1.414)^x$$

(b) Using the same method as before, the initial population (10) is equal to a. The doubling time is 6 minutes, so $y = 20$ when $x = 6$. This is substituted into the exponential model equation to determine b:

$$20 = 10 \times b^6$$

So $b^6 = 2$ and $b = 2^{1/6} \simeq 1.122$, and the exponential model equation is:

$$y = 10 \times (1.122)^x$$

(c) The initial population gives $a = 1000$. The population has halved in 23 minutes, so $y = 500$ when $x = 23$. Substituting into the equation gives:

$$500 = 1000 \times b^{23}$$

So $b = (0.5)^{1/23} \simeq 0.9703$, and the exponential model equation is:

$$y = 1000 \times (0.9703)^x$$

6

(a) If there are a atoms of uranium 239 at the start of an experiment, then after 23.5 minutes there will $a/2$ atoms. So after another 23.5 minutes there will be half of $a/2$ atoms, namely $a/4$ atoms. This is exactly the time period that you are looking for, the time for the number of atoms to reduce by a factor of four. So the 'quarter-life' of the isotope is 47 minutes.

(b) The easiest way to solve this part of the question is to find the equation of the decay first. Let the decay constant be b, then the equation describing the number of uranium atoms is:

$$y = ab^x$$

After 23.5 minutes the number of atoms has halved, so:

$$a/2 = ab^{23.5}$$
$$0.5 = b^{23.5}$$
$$b = (0.5)^{1/23.5}$$
$$b \simeq 0.971$$

Now the 'third-life' can be calculated; it is the time for the number of uranium atoms to decrease to one third. So, substituting $a/3$ for y in the equation

$$y = a \times 0.971^x$$

gives:

$$a/3 = a \times 0.971^x$$
$$\tfrac{1}{3} = 0.971^x$$

Taking logarithms of both sides:

$$\log_{10} \tfrac{1}{3} = \log_{10} 0.971^x$$
$$\log_{10} \tfrac{1}{3} = x \times \log_{10} 0.971$$
$$x = \frac{\log_{10} \frac{1}{3}}{\log_{10} 0.971}$$
$$x \simeq 37 \text{ (to the nearest minute)}$$

So the 'third-life' of uranium 239 is about 37 minutes. (As a rough check, this time should be somewhere between the half-life and the 'quarter-life', which it is.)

7

(a) Let p be the bird population at time t. Using the calculator (and rounding to three significant figures to match the data), gives the best fit exponential model as:

$$p = 555 \times (0.880)^t$$

(Since $r = {}^-0.993$ this model fits the data well and this is confirmed by the graph.)

The model gives a prediction for the population in the year 2005 of $555 \times (0.880)^{15} = 82$ birds (to the nearest whole number).

(b) Using a linear model for the data produces:

$$p = 514 - 40x$$

with coefficients rounded to the nearest whole number. (Since $r = {}^-0.998$ this model is a very good fit.)

The model gives a prediction for the population in the year 2005 of $514 - 40 \times 15 = {}^-86$ birds.

(c) The models predict significantly different numbers of birds. The number of birds predicted by the linear model is negative, which means that this model predicts that the species will be extinct by the year 2005 (it predicts that the extinction will occur about year 12 or 13). However, both models predict that the population is in dire trouble.

8

(a) For Bank A,

$$\text{APR} = (1.009)^{12} \times 100 - 100 = 11.4\% \text{ (correct to 1 d.p.).}$$

For Bank B,

$$\text{APR} = (1.00028)^{365} \times 100 - 100 = 10.8\% \text{ (correct to 1 d.p.).}$$

So, Bank B is offering the better deal.

(b) Use a similar method to that used in Activity 44(ii). The answers below are all given to 3 d.p.

(i) The monthly rate $= 100(\sqrt[12]{1.18} - 1) = 1.389\%$.

(ii) The weekly rate $= 100(\sqrt[52]{1.18} - 1) = 0.319\%$.

(iii) The daily rate $= 100(\sqrt[365]{1.18} - 1) = 0.045\%$.

Unit 13

1

(a) (i) The variables are number of pages and time to read them. Call them p and t (in hours or in minutes if you prefer).

(ii) Assume a constant reading rate of 20 pages per hour (or 3 minutes per page).

(iii) $t = \frac{1}{20}p$ (or $t = 3p$ if you used minutes)

The constant of proportionality is the time to read one page—that is, $\frac{1}{20}$ of an hour (or 3 minutes).

(b) (i) $t = \frac{1}{20} \times 115 = 5.75$ hours or $5\frac{3}{4}$ hours (or $3 \times 115 = 345$ min)

(ii) $t = \frac{1}{20} \times 46 = 2.3$ hours or 138 minutes) That is, nearly 2 hours 20 minutes.

(iii) $t = \frac{1}{20} \times 670 = 33.5$ hours (or $3 \times 670 = 2010$ minutes), i.e. $33\frac{1}{2}$ hours.

Note these answers have been rounded as the model is unlikely to give the prediction very accurately.

2 The depth d is inversely proportional to the square of the diameter of the dish (See Section 1.4, Example 4). So the depth would be scaled by a factor of $(15/20)^2 = 0.5625$, and so the depth of the food in the dish would almost be halved.

Alternatively, the volume using a 15 cm dish would be $V = 7.5^2H = 56.25H$, where H is the height in cm. With a 20 cm dish $V = 10^2h = 100h$, where h is the height so $100h = 56.25H$. So $h = 0.5675H$.

3 If the height of the cakes is to be the same then the volume of the cake is proportional to the area of the base of the cake tin. For a square tin of side x inches, the area is x^2 square inches. For a round tin of diameter d inches, the radius is $d/2$ inches and so the area is $\pi(d/2)^2$ square inches. Equating these two areas gives:

$$x^2 = \pi(d/2)^2 \simeq 0.785\, d^2$$

This relationship can be simplified by taking the square root of both sides to get:

$$x = 0.886\, d$$

For an 8-inch round cake tin, $d = 8$; and so $x = 0.886 \times 8 \simeq 7$. So a 7-inch square tin is equivalent to an 8-inch round tin.

4

(a) $$V = \pi r^2 h$$

Dividing both sides by πh:

$$r^2 = \frac{V}{\pi h}.$$

Taking the square root of both sides:

$$r = \sqrt{\frac{V}{\pi h}}.$$

(b) $$V = \tfrac{4}{3}\pi r^3$$

Dividing both sides by $\frac{4}{3}\pi$:

$$\frac{V}{\frac{4}{3}\pi} = r^3.$$

Rearranging, this gives:

$$\frac{3V}{4\pi} = r^3.$$

Taking the cube root of both sides:

$$r = \sqrt[3]{\frac{3V}{4\pi}}.$$

5

(a) To obtain a simple relationship between cooking times and microwave power you must assume that the amount the food is cooked depends only on the total amount of microwave energy input. This means that:

$$\begin{aligned} \text{amount food is cooked} &\propto \text{energy} \\ &= \text{power rating} \\ &\quad \times \text{cooking time} \end{aligned}$$

So the cooking time is inversely proportional to the power rating of the oven.

(This relationship works well for a given item of food but is not very good for translating cooking times between different items. For different food items other factors come into play: for example, the cooking time also depends on the shape of the food and the proportion of the microwave power absorbed by the food.)

(b) The model is probably valid only over a small range of microwave power, say from about 100 watts to 1000 watts, because it relies on the food responding linearly to the amount of microwave energy input. For small power ratings the approximation will fail because the heat will dissipate by conduction out of the food and the food will not reach a high enough temperature to cook. For very large power ratings the foodstuff's ability to absorb microwaves will be saturated.

(c) If A is the cooking time for the 500-watt oven and B is the cooking time for the 650-watt oven, then from part (a):

$$500A = 650B$$

Therefore:

$$A = \frac{650}{500}B = \frac{13}{10}B$$

and

$$B = \frac{500}{650}A = \frac{10}{13}A$$

These formulas can be used to complete the table as follows.

Food item	Cooking time (minutes)	
	500-watt oven	650-watt oven
One corn-on-the-cob	$6\frac{1}{2}$	5
225 g oven chips	9	7
Whole kipper	8	6
4×225 g cod steaks	$15\frac{1}{2}$	12
2 kg beef	64	49
2×100 g pork chops	6	$4\frac{1}{2}$

6

(a) Since you are told that both screens are the traditional shape, the television screens must be mathematically similar figures and one is just a scaled version of another. So the 21-inch screen is scaled up by 24/21 to get the 24-inch screen. So the area is scaled up by this factor squared: $(24/21)^2 \simeq 1.31$. So the 24-inch television would be perceived to be 1.31 times, or 31%, bigger than the 21-inch one.

(b) The argument in the previous part applies. This gives the following equation:

$$\text{area of large television} = \left(\frac{x}{21}\right)^2 \times \text{area of 21-inch television}$$

So the perceived increase in size will be:

$$\text{increase} = \left(\frac{x}{21}\right)^2$$

It is this quadratic increase in perceived size which makes televisions with only a slightly larger diagonal *seem* much larger.

7

(a) The surface area of the sweets is proportional to r^2 and the volume to r^3. So as r goes up by a factor of 3 the surface coating will go up by a factor of $3^2 = 9$ and the interior will go up by a factor of $3^3 = 27$.

(b) The weight of the sweets will scale in exactly the same proportion as the volume. So one giant sweet will weigh 27 times as much as one small sweet, or to put it another way one small sweet will weigh 1/27 as much as a giant sweet. So 600 small sweets will weigh the same as $600/27 \simeq 22$ giant sweets. So 22 giant sweets would need to be put in a packet.

(c) Each giant sweet needs 9 times as much coating, but there are now 22 sweets in a packet instead of 600. So for a packet of giant sweets the coating ingredients would have to be scaled by $9 \times 22/600 = 0.33$.

For the interior of the sweets, each giant sweet needs 27 times more ingredients, but there are now 22 sweets in a packet instead of 600. So for a packet, the interior ingredients would have to be scaled by $27 \times 22/600 = 0.99$. So the amount of coating is reduced to a third (0.33) but the amount of ingredients for the core is virtually the same.

(d) The scaling for 100 packets would be the same as the scaling for one packet. So the answer remains the same: the interior ingredients would be roughly unchanged (scaling factor 0.99) but the coating ingredients would need to be scaled down to about a third (scaling factor 0.33).

8

(a) Let t be the time taken to dig the hole. The volume of the hole to be dug is $x \times x \times x/2 = x^3/2$. Assume that the amount of time taken to dig a hole is directly proportional to the volume of soil that has to be moved, so

$$t = k \times x^3/2$$

where k is a constant. To determine k, use the data from the test hole. When $x = 1\,\text{m}$, $t = 2$ hours. Substitute this into the equation:

$$2 = k \times 1^3/2.$$

This gives $k = 4$. The formula for the time taken is then as follows:

$$t = 2 \times x^3.$$

(b) For this case $x = 5$, so:

$$t = 2 \times 5^3 = 250 \text{ hours.}$$

So the time taken to dig the whole swimming pool is 250 hours. (Based on this large estimate for the time, it might be more sensible to pay to hire an excavator!)

9

(a) I is the energy per unit area per second; so, to find the total energy per second, multiply by the area:

$$\begin{aligned}\text{total energy} &= \text{intensity} \times \text{area}\\ &= I \times 4\pi r^2\end{aligned}$$

(b) The total energy per second calculated in the previous part of the question must equal the energy emitted by the star per second. Therefore:

$$E = I \times 4\pi r^2$$

This gives the relationship between intensity and distance as:

$$I = \frac{E}{4\pi r^2} \text{ or } I = \frac{E}{4\pi} \times r^{-2}$$

This is an inverse square proportionality relationship and is known in physics as an *inverse square law*. Such laws often occur in physics for exactly the same reasons as above (that is, something is being distributed equally in all directions).

(To use the above formula, the amount of light energy E emitted by the star would have to be known. One way of determining this is to use a relationship between E and the colour of the light emitted. The colour of the light emitted by a star is often used to help describe the star: for example, red giant, brown dwarf, and so on.)

10 The correct pairings are 1e, 2d, 3c, 4b, 5a and 6f.